METHODS IN MOLECULAR BIOLOGY

Series Editor
John M. Walker
School of Life and Medical Sciences
University of Hertfordshire
Hatfield, Hertfordshire, UK

For further volumes:
http://www.springer.com/series/7651

For over 35 years, biological scientists have come to rely on the research protocols and methodologies in the critically acclaimed *Methods in Molecular Biology* series. The series was the first to introduce the step-by-step protocols approach that has become the standard in all biomedical protocol publishing. Each protocol is provided in readily-reproducible step-by-step fashion, opening with an introductory overview, a list of the materials and reagents needed to complete the experiment, and followed by a detailed procedure that is supported with a helpful notes section offering tips and tricks of the trade as well as troubleshooting advice. These hallmark features were introduced by series editor Dr. John Walker and constitute the key ingredient in each and every volume of the *Methods in Molecular Biology* series. Tested and trusted, comprehensive and reliable, all protocols from the series are indexed in PubMed.

Plant Antiviral Resistance

Methods and Protocols

Edited by

Guanwei Wu

Institute of Plant Virology, Ningbo University, Ningbo, China

Xiaofei Cheng

College of Plant Protection, Northeast Agricultural University, Harbin, China

Humana Press

Editors
Guanwei Wu
Institute of Plant Virology
Ningbo University
Ningbo, China

Xiaofei Cheng
College of Plant Protection
Northeast Agricultural University
Harbin, China

ISSN 1064-3745 ISSN 1940-6029 (electronic)
Methods in Molecular Biology
ISBN 978-1-0716-5299-2 ISBN 978-1-0716-5300-5 (eBook)
https://doi.org/10.1007/978-1-0716-5300-5

This Humana imprint is published by the registered company Springer Science+Business Media, LLC, part of Springer Nature.
The registered company address is: 1 New York Plaza, New York, NY 10004, U.S.A.

Preface

In an era of growing global food demands and the ongoing challenge of emerging plant pathogens, the study of plant antiviral resistance has become a vital area of agricultural science. Viral diseases continue to cause significant crop losses worldwide, threatening food security and livelihoods. However, plants possess remarkable defense mechanisms that can evolve and counteract viral threats—offering both hope and a rich avenue for scientific exploration.

This volume aims to provide a comprehensive collection of modern experimental techniques, protocols, and analytical tools for studying plant antiviral resistance. It covers a wide range of topics, from the molecular mechanisms of viral infection and plant immunity (e.g., RNA interference, autophagy, immune signaling, and viral reverse genetics) to advanced methodologies for high-throughput screening, omics-based profiling (metabolomics, proteomics), and cutting-edge proximity tagging technology applications. Additionally, it explores practical strategies for developing virus-resistant crops, including cross-protection, transgenic approaches, shoot-tip culture, and gene editing.

We are deeply grateful to the authors who have contributed their valuable protocols and insights to this volume. We apologize to any researchers whose relevant work may not have been cited and to readers for any omissions regarding plant antiviral resistance methods.

Finally, I hope this volume will serve as a valuable resource for undergraduate students and researchers working in the field of plant antiviral resistance, offering both foundational knowledge and cutting-edge approaches to address this critical issue.

Ningbo, China *Guanwei Wu*
Harbin, China *Xiaofei Cheng*

Contents

Contributors

SHUAIFA CHANG • *College of Plant Protection, Northeast Agricultural University, Harbin, Heilongjiang, P. R. China*

JIANPING CHEN • *State Key Laboratory for Quality and Safety of Agro-products, Key Laboratory of Biotechnology in Plant Protection of MARA, Key Laboratory of Green Plant Protection of Zhejiang Province, Institute of Plant Virology, Ningbo University, Ningbo, China*

XIAOFEI CHENG • *College of Plant Protection, Northeast Agricultural University, Harbin, Heilongjiang, P. R. China*

HONGGUANG CUI • *Key Laboratory of Green Prevention and Control of Tropical Plant Diseases and Pests (Ministry of Education), School of Tropical Agriculture and Forestry, Hainan University, Haikou, Hainan, China*

XINYUN CUI • *College of Plant Science, Jilin University, Changchun, Jilin, P. R. China*

ZHAOJI DAI • *Key Laboratory of Green Prevention and Control of Tropical Plant Diseases and Pests (Ministry of Education), School of Tropical Agriculture and Forestry, Hainan University, Haikou, Hainan, China*

CHANGHUI DENG • *Key Laboratory of Green Prevention and Control of Tropical Plant Diseases and Pests (Ministry of Education) and School of Tropical Agriculture and Forestry, Hainan University, Haikou, China*

YAFENG DONG • *Research Institute of Pomology, Chinese Academy of Agricultural Sciences, Xingcheng, China*

XINYUE FAN • *College of Plant Protection, Northeast Agricultural University, Harbin, Heilongjiang, P. R. China*

XINXIN FANG • *State Key Laboratory for Quality and Safety of Agro-products, Key Laboratory of Biotechnology in Plant Protection of MARA, Key Laboratory of Green Plant Protection of Zhejiang Province, Institute of Plant Virology, Ningbo University, Ningbo, China*

RUIFANG GAO • *College of Plant Science, Jilin University, Changchun, Jilin, P. R. China*

ZHONGXIN GUO • *College of Life Sciences, Shandong Agricultural University, Tai'an, Shandong, China*

KELEI HAN • *Institute of Plant Protection and Agro-Products Safety, Anhui Academy of Agricultural Sciences, Hefei, Anhui, China*

NI HONG • *Key Lab of Plant Pathology of Hubei Province, College of Plant Science and Technology, Huazhong Agricultural University, Wuhan, China*

GUOJUN HU • *Research Institute of Pomology, Chinese Academy of Agricultural Sciences, Xingcheng, China*

YIXIN HUANG • *College of Plant Protection, Northeast Agricultural University, Harbin, Heilongjiang, P. R. China*

KAZUHIRO ISHIBASHI • *Institute of Agrobiological Sciences, National Agriculture and Food Research Organizations (NARO), Tsukuba, Ibaraki, Japan*

MASAYUKI ISHIKAWA • *Institute of Agrobiological Sciences, National Agriculture and Food Research Organizations (NARO), Tsukuba, Ibaraki, Japan*

DANGCHENG JI • *College of Plant Protection, Northeast Agricultural University, Harbin, Heilongjiang, P. R. China*

WEIQIN JI • *College of Plant Protection, Northeast Agricultural University, Harbin, Heilongjiang, P. R. China*
XIAOYE KE • *Pepper Germplasm Research Group, College of Agriculture, Anhui Science and Technology University, Chuzhou, Anhui, P. R. China*
KAIJING LI • *College of Plant Protection, Northeast Agricultural University, Harbin, Heilongjiang, P. R. China*
NA LI • *Key Laboratory of Green Prevention and Control of Tropical Plant Diseases and Pests (Ministry of Education) and School of Tropical Agriculture and Forestry, Hainan University, Haikou, China*
YONG LI • *College of Life Science, Northeast Agricultural University, Harbin, Heilongjiang, P. R. China*
JIAJIA LIN • *State Key Laboratory for Quality and Safety of Agro-products, Key Laboratory of Biotechnology in Plant Protection of MARA, Key Laboratory of Green Plant Protection of Zhejiang Province, Institute of Plant Virology, Ningbo University, Ningbo, China*
JIAYU LIN • *The Engineering Research Center for Plant Health Protection Technology in Henan Province, College of Plant Protection, Henan Agricultural University, Zhengzhou, China*
JIA LIU • *College of Plant Protection, Northeast Agricultural University, Harbin, Heilongjiang, P. R. China*
WENCHENG LIU • *Shanghai Collaborative Innovation Center of Agri-Seeds/School of Agriculture and Biology, Shanghai Jiao Tong University, Shanghai, China*
CHAO MA • *Institute of Plant Protection and Agro-Products Safety, Anhui Academy of Agricultural Sciences, Hefei, Anhui, China*
WENQI MAO • *State Key Laboratory for Quality and Safety of Agro-products, Key Laboratory of Biotechnology in Plant Protection of MARA, Key Laboratory of Green Plant Protection of Zhejiang Province, Institute of Plant Virology, Ningbo University, Ningbo, China*
JIANCHUN QIN • *College of Plant Science, Jilin University, Changchun, Jilin, P. R. China*
PENGHUAN RUI • *State Key Laboratory for Quality and Safety of Agro-Products, Key Laboratory of Biotechnology in Plant Protection of MARA, Key Laboratory of Green Plant Protection of Zhejiang Province, Institute of Plant Virology, Ningbo University, Ningbo, China*
YAN SHI • *College of Plant Protection, Henan Agricultural University, Zhengzhou, Henan, China*
BO SONG • *College of Plant Protection, Northeast Agricultural University, Harbin, Heilongjiang, P. R. China*
CHENGGANG SONG • *College of Plant Science, Jilin University, Changchun, Jilin, P. R. China*
BINGJIAN SUN • *The Engineering Research Center for Plant Health Protection Technology in Henan Province, College of Plant Protection, Henan Agricultural University, Zhengzhou, China*
HANGJUN SUN • *The Engineering Research Center for Plant Health Protection Technology in Henan Province, College of Plant Protection, Henan Agricultural University, Zhengzhou, China*
ADRIAN VALLI • *Centro Nacional de Biotecnología, CSIC, Madrid, Spain*
GUOPING WANG • *Key Lab of Plant Pathology of Hubei Province, College of Plant Science and Technology, Huazhong Agricultural University, Wuhan, China*
HONGRU WANG • *College of Plant Protection, Northeast Agricultural University, Harbin, Heilongjiang, P. R. China*

LINXI WANG • *Key Laboratory of Green Prevention and Control of Tropical Plant Diseases and Pests (Ministry of Education), School of Tropical Agriculture and Forestry, Hainan University, Haikou, Hainan, China*
KIKYO WATANABE • *Institute of Agrobiological Sciences, National Agriculture and Food Research Organizations (NARO), Tsukuba, Ibaraki, Japan*
GUANWEI WU • *State Key Laboratory for Quality and Safety of Agro-products, Key Laboratory of Biotechnology in Plant Protection of MARA, Key Laboratory of Green Plant Protection of Zhejiang Province, Institute of Plant Virology, Ningbo University, Ningbo, China*
XIAOYUN WU • *College of Plant Protection, Northeast Agricultural University, Harbin, Heilongjiang, P. R. China*
DANKAN YAN • *Institute of Plant Protection and Agro-Products Safety, Anhui Academy of Agricultural Sciences, Hefei, Anhui, China*
FEI YAN • *State Key Laboratory for Quality and Safety of Agro-products, Key Laboratory of Biotechnology in Plant Protection of MARA, Key Laboratory of Green Plant Protection of Zhejiang Province, Institute of Plant Virology, Ningbo University, Ningbo, China*
WENKAI YAN • *State Key Laboratory for Quality and Safety of Agro-Products, Key Laboratory of Biotechnology in Plant Protection of MARA, Key Laboratory of Green Plant Protection of Zhejiang Province, Institute of Plant Virology, Ningbo University, Ningbo, China*
LINGLING YANG • *College of Plant Protection, Henan Agricultural University, Zhengzhou, Henan, China*
XUE YANG • *College of Plant Protection, Henan Agricultural University, Zhengzhou, Henan, China*
TETSUYA YOSHIDA • *Institute of Agrobiological Sciences, National Agriculture and Food Research Organizations (NARO), Tsukuba, Ibaraki, Japan*
LIANSHENG YU • *College of Plant Protection, Northeast Agricultural University, Harbin, Heilongjiang, P. R. China*
HONGTAO ZHANG • *The Engineering Research Center for Plant Health Protection Technology in Henan Province, College of Plant Protection, Henan Agricultural University, Zhengzhou, China*
WEI ZHANG • *Qianxinan Zhou Company of Guizhou Provincial Tobacco Corporation, Qianxinan, China*
XIANGLONG ZHANG • *Pepper Germplasm Research Group, College of Agriculture, Anhui Science and Technology University, Chuzhou, Anhui, P. R. China*
XIANGXIANG ZHANG • *Pepper Germplasm Research Group, College of Agriculture, Anhui Science and Technology University, Chuzhou, Anhui, P. R. China*

Chapter 1

Development of Tobamovirus-Resistant Tomato Plants by CRISPR-Cas9-Mediated Knockout of Susceptibility Genes

Kikyo Watanabe, Masayuki Ishikawa, and Kazuhiro Ishibashi

Abstract

The advent of genome editing technologies such as CRISPR-Cas9 has revolutionized the development of disease-resistant crops, offering precision and efficiency in targeting specific genetic loci responsible for susceptibility. In this protocol, we harness the CRISPR-Cas9 system to disrupt key susceptibility genes in tomato, aiming to fortify resistance against tobamoviruses, particularly the aggressive tomato brown rugose fruit virus (ToBRFV). By systematically knocking out four *TOM1* homologs, genes essential for tobamoviral replication, tomato lines with robust and heritable resistance can be developed while minimizing adverse developmental effects. The approach not only underscores the significance of basic research on host—pathogen interactions in modern crop protection but also lays the groundwork for sustainable, gene-driven resistance strategies in commercial tomato breeding.

Key words Tomato, CRISPR-Cas9, Tobamovirus, ToBRFV, *TOM1*, Genome editing, Disease resistance, Susceptibility genes, Sustainable agriculture

1 Introduction

Tomato brown rugose fruit virus (ToBRFV) is a recently emerged, devastating virus that affects tomato cultivation worldwide [1, 2]. ToBRFV belongs to the genus *Tobamovirus* within the family *Virgaviridae*, and is closely related to tomato mosaic virus (ToMV). Although ToMV was previously a major threat to tomato production, it has been effectively controlled through the introduction of the resistance gene *Tm-2*^{2} [3]. However, *Tm-2*^{2} does not confer resistance to ToBRFV, allowing the virus to spread rapidly via international transport of contaminated seeds. Therefore, new genetic resources for ToBRFV resistance are urgently needed.

Viruses possess small genomes with limited coding capacity and rely heavily on host factors for their multiplication [4, 5]. Loss-of-function mutations in genes encoding these factors can confer recessive resistance to the corresponding viruses [6]. These

Guanwei Wu and Xiaofei Cheng (eds.), *Plant Antiviral Resistance: Methods and Protocols*, Methods in Molecular Biology, vol. 3039,
https://doi.org/10.1007/978-1-0716-5300-5_1,

Table 1
***TOM1* homologs in the tomato genome**

TOM1 homologs	Gene ID
SlTOM1a	Solyc04g008540
SlTOM1b	Solyc01g105270
SlTOM1c	Solyc02g080370
SlTOM1d	Solyc01g007900

"susceptibility genes" are promising targets for resistance breeding. However, successful cases are limited, as these genes often play essential roles in plant development, and their disruption may lead to undesirable phenotypes.

Several susceptibility genes for tobamoviruses have been identified [7, 8]. *Tobomovirus multiplication 1* (*TOM1*) was discovered through forward genetic screening of *Arabidopsis thaliana* mutants with impaired tobamovirus multiplication [9, 10]. In *Arabidopsis*, replication of tobamoviruses is undetectable in protoplasts of the double mutant of *TOM1* and its paralog *TOM3*, while the mutant does not exhibit obvious detrimental effects on plant vigor [11, 12]. Given these characteristics and the conservation of *TOM1* homologs across Viridiplantae, *TOM1* is considered a good target for resistance breeding against tobamoviruses. Indeed, RNA interference-mediated knockdown of *TOM1* homologs in tobacco confers resistance to several tobamoviruses [13].

In the tomato genome, four expressed *TOM1* homologs (*SlTOM1a–d*) have been identified (Table 1), complicating the use of *TOM1* for resistance breeding. Recent advances in biotechnology, such as CRISPR-Cas9-mediated genome editing, enable simultaneous disruption of multiple genes. Here, we describe a method to produce tobamovirus-resistant tomato plants by knocking out *SlTOM1a–d*. While triple mutants show reduced ToBRFV multiplication, quadruple mutants exhibit enhanced resistance [14].

2 Materials

2.1 Plant and Bacterial Strains

1. Tomato (*Solanum lycopersicum*) cultivar Craigella GCR26.
2. *Escherichia coli* for standard transformation.
3. *Rhizobium radiobacter* (*Agrobacterium*) GV3101::pMP90.

2.2 Plasmids

1. pDeCas9 [15].
2. pDeCas9-SlTOM1 (Fig. 1), a derivative of pDeCas9 constructed for *SlTOM1a–d* disruption [14].

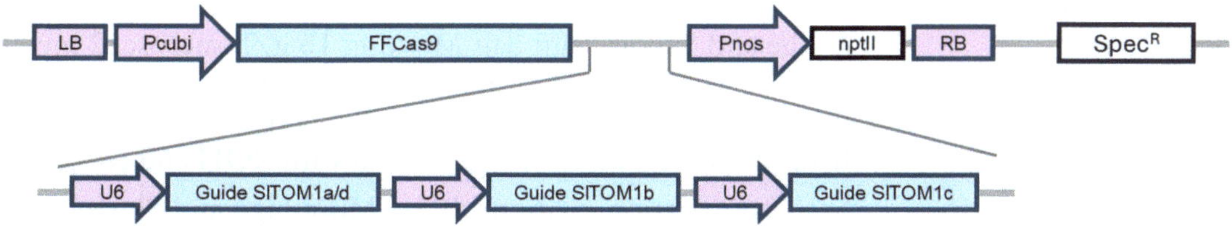

Fig. 1 Structure of pDeCas9-SlTOM1

2.3 Reagents and Miscellaneous

1. Standard DNA polymerase.
2. In vitro transcription reagent.
3. PCR reagents and thermocycler.
4. Genomic DNA extracted from tomato.
5. Cas9 Nuclease, *Streptococcus pyogenes* and a suitable buffer.
6. Standard agarose gel electrophoresis reagents and systems.
7. Sterile water.
8. 70% ethanol.
9. Kitchen Haiter (Kao).
10. Fine scalpel.
11. Autoclaved paper towels.
12. 100 mM acetosyringone stock solution prepared in dimethyl sulfoxide.
13. 2-Mercaptoethanol.
14. Spectinomycin and kanamycin for bacterial and plant selection, respectively.
15. Timentin for *Agrobacterium* elimination.
16. LB medium.
17. Centrifuge for collecting bacteria.
18. Spectrophotometer (OD_{600}).
19. Gas-permeable surgical tape.
20. Plastic petri dishes with $\phi 90 \times 20$ mm.
21. PHYTATRAY II (Sigma).
22. Pots, pre-wetted soil, plastic wrap for acclimatization.
23. Growth chamber for cultivating tomato plants under a 16-h light/8-h dark cycle at 25 °C.
24. 10 mM Sodium phosphate buffer (NaPi) pH 7.0 for virus inoculation.
25. Carborundum.
26. Disposable gloves.

2.4 Media for Tomato Tissue Culture

1. MS liquid medium for *Agrobacterium* inoculation: 0.5× Murashige and Skoog Plant Salt Mixture, 15 g/L sucrose, 2 mg/L glycine, 0.1 g/L myo-inositol, 0.5 mg/L nicotinic acid, 0.5 mg/L pyridoxine hydrochloride, and 0.1 mg/L thiamin hydrochloride. Adjust the pH to 5.8 with KOH and autoclave.
2. Seed germination medium: 0.5× Murashige and Skoog Plant Salt Mixture, 15 g/L sucrose, 2 mg/L glycine, 0.1 g/L myo-inositol, 0.5 mg/L nicotinic acid, 0.5 mg/L pyridoxine hydrochloride, and 0.1 mg/L thiamin hydrochloride. Adjust the pH to 5.8 with KOH and add 3 g/L gellan gum. After autoclaving, pour the medium (ca. 40 mL per tray) into plant boxes (PHYTATRAY II, Sigma) and solidify them.
3. Co-cultivation medium: 1× Murashige and Skoog Plant Salt Mixture, 30 g/L sucrose, 2 mg/L glycine, 0.1 g/L myo-inositol, 0.5 mg/L nicotinic acid, 0.5 mg/L pyridoxine hydrochloride, 0.1 mg/L thiamin hydrochloride, and 1.5 mg/L trans-Zeatin. Adjust the pH to 5.8 with KOH and add 3 g/L gellan gum. After autoclaving, pour the medium (ca. 40 mL per plate) into petri dishes (ϕ90 × 20 mm) and solidify them.
4. Callus induction medium: 1× Murashige and Skoog Plant Salt Mixture, 30 g/L sucrose, 2 mg/L glycine, 0.1 g/L myo-inositol, 0.5 mg/L nicotinic acid, 0.5 mg/L pyridoxine hydrochloride, 0.1 mg/L thiamin hydrochloride, and 1.5 mg/L trans-Zeatin. Adjust the pH to 5.8 with KOH and add 3 g/L gellan gum. After autoclaving, add timentin and kanamycin to 0.15 g/L and 0.1 g/L, respectively, and pour the medium (ca. 40 mL per plate) into petri dishes (ϕ90 × 20 mm) and solidify them.
5. Shoot elongation medium: 1× Murashige and Skoog Plant Salt Mixture, 30 g/L sucrose, 2 mg/L glycine, 0.1 g/L myo-inositol, 0.5 mg/L nicotinic acid, 0.5 mg/L pyridoxine hydrochloride, 0.1 mg/L thiamin hydrochloride, and 1 mg/L trans-Zeatin. Adjust the pH to 5.8 with KOH and add 3 g/L gellan gum. After autoclaving, add timentin and kanamycin to 0.15 g/L and 0.1 g/L, respectively, and pour the medium (ca. 40 mL per plate) into petri dishes (ϕ90 × 20 mm) and solidify them.
6. Root induction medium: 0.5× Murashige and Skoog Plant Salt Mixture, 15 g/L sucrose, 2 mg/L glycine, 0.1 g/L myo-inositol, 0.5 mg/L nicotinic acid, 0.5 mg/L pyridoxine hydrochloride, and 0.1 mg/L thiamin hydrochloride. Adjust the pH to 5.8 with KOH and add 3 g/L gellan gum. After autoclaving, add timentin and kanamycin to 0.15 g/L and 0.1 g/L, respectively, and pour the medium (ca. 40 mL per plate or tray) into petri dishes (ϕ90 × 20 mm) or PHYTATRAY II plant boxes and solidify them.

Table 2
Oligonucleotides used for *SlTOM1c* target cleavage assay

Oligo name	Sequence
SpCas9-scaffold-R	AAAAAAAAGCACCGACTCGGTGCCACTTTTTCAAGTTGATAACGGACTAG CCTTATTTTAACTTGCTATTTCTAGCTCTAAAAC
T7-SlTOM1c-F1 (for Sg-1)	GATCAC**TAATACGACTCACTATA***GCCGATTGACATCGCCGGTC*GTTTTAGA GCTAGAAATAGCAA
T7-SlTOM1c-F2 (for Sg-2)	GATCAC**TAATACGACTCACTATAG***TCGCCGGTCCGGTGACCAAC*GTTTTAG AGCTAGAAATAGCAA
SlTOM1c_G1F	CGCGTTAAATTCGTACATGATC
SlTOM1c_G1R	ATACTCGGGTACCCTCAAATCA

Red, bold, and italicized characters represent Cas9 scaffold, T7 RNA polymerase promoter and guide sequences, respectively

2.5 Oligonucleotides Used for* SlTOM1c *Target Cleavage Assay

1. Listed in Table 2.

2.6 Softwear/ Resources

1. CRISPR-P program (http://cbi.hzau.edu.cn/crispr/; [16]).

3 Methods

3.1 Design of sgRNA for the CRISPR/Cas9 System

1. Run the CRISPR-P program, and obtain candidate guide RNA (gRNA) sequences. Their scores for offtarget sites should be low.
2. Test the top two or more candidates for the activity to cleave target DNA in vitro. An example for *SlTOM1c* is described below. To prepare the DNA template for single guide (sg) RNA transcription, anneal the synthetic oligonucleotides SpCas9-Scaffold-R with either T7-SlTOM1c-F1 or T7-SlTOM1c-F2 (as listed in Table 2). Subsequently, generate the double-stranded DNA by primer extension using a standard DNA polymerase. From these double-stranded DNA fragments, synthesize sgRNA using an in vitro transcription reagent such as AmpliScribe T7-Flash transcription kit (CELLSCRIPT). Prepare a target

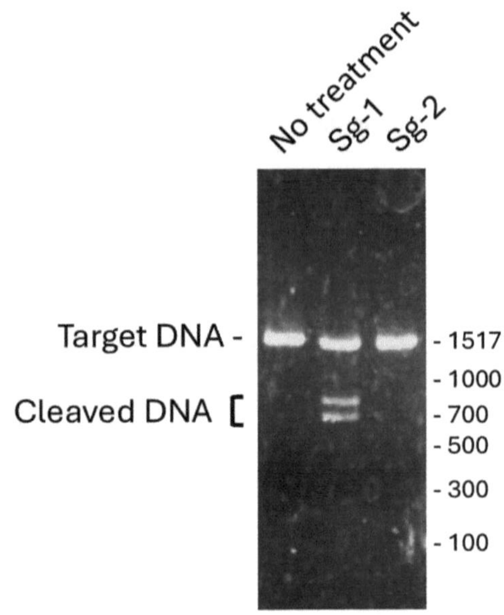

Fig. 2 An example of in vitro cleavage of a target DNA fragment by sgRNA-Cas9 complexes. *SlTOM1c* guide RNA candidates Sg-1 and Sg-2 were tested

Table 3
Nucleotide sequences of the guide RNA

Target *TOM1* homologs	Guide RNA sequence
SlTOM1a and *d*	GUUGUGAAGAAUGCAAGACU
SlTOM1b	UCAAAGAUGGGGCGAACUCG
SlTOM1c	GCCGAUUGACAUCGCCGGUC

DNA fragment by PCR with synthetic oligonucleotide primers SlTOM1c_G1F and SlTOM1c_G1R (Table 2) using tomato genomic DNA as a template.

3. Assemble a sgRNA-Cas9 complex by mixing 1 pmol sgRNA and 1 pmol Cas9 in 30 μL of 1× Cas9 nuclease reaction buffer and by incubating at 37 °C for 10 min. Mix 9 μL of the reaction mixture with 0.2 pmol (1 μL) of the target DNA fragment, incubate at 37 °C for 1 h, and analyze the products by agarose gel electrophoresis (Fig. 2). sgRNA sequences with higher target cleavage activity (in Fig. 2 case, Sg-1) are shown in Table 3.

3.2 Inactivation of SlTOM1 Genes

Transformation of tomato is essentially as described by Sun et al. [17]. Culture tomato plants and tomato explant-derived tissues at 25 °C under a 16-h light (ca. 5000 lux)/8-h dark cycle in a growth chamber unless otherwise stated.

1. Construct a plasmid carrying Cas9- and sgRNA-expressing gene cassettes and a gene for antibiotic selection of transformed plants in the T-DNA region (pDeCas9-SlTOM1 is shown as an example in Fig. 1).

2. Sterilize tomato seeds (Craigella GCR26 or other cultivars susceptible to tomato mosaic virus and several other tobamoviruses) by incubating in 70% ethanol for 2 min with agitation, washing with sterile water 3 times, and incubating in tenfold-diluted commercial kitchen bleach containing sodium hypochlorite and detergent (Kitchen Haiter, Kao, Tokyo, Japan) for 45 min at room temperature. Wash the seeds with sterile water several times, incubate them overnight at 30 °C in sterile water in the dark. Apply 2 mL of sterile water on germination medium to maintain moist conditions prior to germination and then sow 20—25 seeds per tray. Incubate the seeds in a growth chamber for 6—8 days.

3. Grow *Agrobacterium* carrying pDeCas9-SlTOM1 (Fig. 1) in 2 mL of LB medium containing 50 μg/mL spectinomycin at 30 °C overnight with vigorous shaking. Collect the bacterial cells by centrifugation and resuspend them in MS liquid medium containing 100 μM acetosyringone and 10 μM 2-mercaptoethanol to OD_{600} 0.07.

4. Prepare cotyledon explants from the 6- to 8-day-old tomato seedlings by cutting in MS liquid medium containing 10 μM 2-mercaptoethanol using a sterilized fine scalpel (Fig. 3) and incubate them in the *Agrobacterium* suspension prepared in Subheading 3.2, **step 3** for 10 min at room temperature (*see* **Note 1**). Gently scoop the leaf explants and blot on autoclaved paper towels to remove excess liquid. Immediately (do not let

Fig. 3 Tomato seedlings for the preparation of explants for transformation. Black lines indicate approximate cutting points. Discard the stem and the leaf apexes

them dry) place the explants on co-cultivation medium with the adaxial surface of the leaf in contact with the medium (Note: Sun et al. [17] let the abaxial surface contact with the medium). Seal plates with gas-permeable surgical tape and incubate them at 25 °C in the dark for 3 days.

5. Transfer the explants to Callus induction medium. Place the leaf explants with the abaxial side in contact with the medium (*see* **Note 2**). Incubate them in a growth chamber. Maintaining cultures under dim light conditions (ca. 1500 lux) for the first week, followed by transfer to bright light conditions (ca. 5000 lux), can facilitate shoot induction. Calli starts to appear around 2 weeks after incubation on Callus induction medium frequently from leaf-stalk regions of the cut surface. Transfer the explants-calli to fresh Callus induction medium every 2 weeks (*see* **Note 3**).
6. When calli reach a diameter of 2—5 mm or adventitious buds develop from calli or leaf explants, transfer them to fresh shoot elongation medium and subculture every 2 weeks (*see* **Note 4**). When the shoots reach 1—2 cm in height, excise stems using a sterilized fine scalpel, ensuring that each stem retains longer than 5 mm between the cut end and the first node). Insert stems at a 45-degree angle into root induction medium so that their basal end touch plate bottom. Transfer the remaining calli with small buds to fresh shoot elongation medium. For larger shoots, substitute the Petri dish lid with an inverted base to provide sufficient vertical space for shoot elongation.
7. Identify genome-edited shoots by cleaved amplified polymorphic sequence (CAPS) analysis, T7 endonuclease I (T7EI) assay, fragment analysis, high-resolution melting curve analysis, or other techniques. Confirm the mutations by sequencing. Although the likelihood of generating shoots with quadruple, bi-allelic loss-of-function mutations is relatively low due to the frequent occurrence of deletions in multiples of three bases, the challenges can be addressed by crossing regenerated plants exhibiting the desired mutations and stacking them as outlined in Subheading 3.3.
8. When the plantlets reach 2—4 cm in height with sufficient root system (at least 3—5 cm in length and 3—5 main roots), place the regenerated shoots in hardening conditions to promote acclimatization. Remove the plantlets from the plates and gently rinse off the adhering gel. Transfer them to pots containing pre-wetted soil and immediately cover with plastic wrap to prevent desiccation. After keeping them in a growth chamber for 7 days, gradually perforate the plastic wrap to reduce sudden shock. Subsequently, incubate them in a growth chamber for 15—20 days, and transfer to a containment greenhouse, and harvest seeds (T1 generation).

Fig. 4 The *sltom1a–d* quadruple mutant tomato plants exhibit strong resistance to ToBRFV

3.3 Stacking* sltom1a–d *Mutations and Virus Inoculation Assay

Cross obtained mutant tomato plants with parental wild-type (GCR26) plant or other mutants, or self-pollinate the obtained mutants, and select progeny with desired genotypes. Eventually, all combinations of single, double, triple, and quadruple *Sltom1a–d* mutant lines can be established. An example of ToBRFV resistance assay in a *sltom1a–d* quadruple mutant line is shown in Fig. 4. For mechanical inoculation of ToBRFV, apply leaf homogenates of infected leaves in 10 mM NaPi buffer (pH 7.0) and abrasive carborundum onto the cotyledons of young seedlings and then rub gently by hand while wearing disposable globes.

4 Notes

1. Callus-forming competence is confined to a short window in the course of cotyledon development. Efficient regeneration is best achieved using cotyledon explants from seedlings at an early stage, when the first true leaf is under 5 mm in length. Change the scalpel blade frequently for fine cutting, because tearing or damage results in low regeneration efficiency.
2. High plantlet density reduces the growth rate. Ensure adequate spacing between explants; typically, we place 15 explants per plate.
3. Wet conditions of the media promote *Agrobacterium* overgrowth. If plates are wet with many water droplets, remove them by pipetting and air dry in a clean bench. To avoid *Agrobacterium* overgrowth, frequently monitor the plates. When signs of *Agrobacterium* overgrowth are observed, wash the explants several times with sterile water and place them onto a new medium.
4. Choose shoots that display normal phyllotaxy with apical meristem. Discard glassy shoots.

Acknowledgement

This work was supported in part by the Research and Implementation Promotion Program through Open Innovation (JPJ011937) from the Project of the Bio-oriented Technology Research Advancement Institution (BRAIN).

Competing Interest Statement M.I. and K.I. are named as inventors on a patent application related to this work.

References

1. Zhang S, Griffiths JS, Marchand G, Bernards MA, Wang A (2022) Tomato brown rugose fruit virus: an emerging and rapidly spreading plant RNA virus that threatens tomato production worldwide. Mol Plant Pathol 23:1262–1277
2. Salem NM, Jewehan A, Aranda MA, Fox A (2023) Tomato brown rugose fruit virus pandemic. Annu Rev Phytopathol 61:137–164
3. Ishibashi K, Kubota K, Kano A, Ishikawa M (2023) Tobamoviruses: old and new threats to tomato cultivation. J Gen Plant Pathol 89: 305–321
4. Nagy PD, Pogany J (2012) The dependence of viral RNA replication on co-opted host factors. Nat Rev Microbiol 10:137–149
5. Wang A (2015) Dissecting the molecular network of virus-plant interactions: the complex roles of host factors. Annu Rev Phytopathol 53:45–66
6. Garcia-Ruiz H (2018) Susceptibility genes to plant viruses. Viruses 10:484
7. Ishibashi K, Miyashita S, Katoh E, Ishikawa M (2012) Host membrane proteins involved in the replication of tobamovirus RNA. Curr Opin Virol 2:693–698
8. Ishibashi K, Ishikawa M (2016) Replication of tobamovirus RNA. Annu Rev Phytopathol 54: 55–78
9. Ishikawa M, Obata F, Kumagai T, Ohno T (1991) Isolation of mutants of Arabidopsis thaliana in which accumulation of tobacco mosaic virus coat protein is reduced to low levels. Mol Gen Genet 230:33–38
10. Ishikawa M, Naito S, Ohno T (1993) Effects of the tom1 mutation of Arabidopsis thaliana on the multiplication of tobacco mosaic virus RNA in protoplasts. J Virol 67:5328–5338
11. Yamanaka T, Imai T, Satoh R, Kawashima A, Takahashi M, Tomita K, Kubota K, Meshi T, Naito S, Ishikawa M (2002) Complete inhibition of tobamovirus multiplication by simultaneous mutations in two homologous host genes. J Virol 76:2491–2497
12. Fujisaki K, Ravelo GB, Naito S, Ishikawa M (2006) Involvement of THH1, an Arabidopsis thaliana homologue of the *TOM1* gene, in tobamovirus multiplication. J Gen Virol 87: 2397–2401
13. Asano M, Satoh R, Mochizuki A, Tsuda S, Yamanaka T, Nishiguchi M, Hirai K, Meshi T, Naito S, Ishikawa M (2005) Tobamovirus-resistant tobacco generated by RNA interference directed against host genes. FEBS Lett 579:4479–4484
14. Ishikawa M, Yoshida T, Matsuyama M, Kouzai Y, Kano A, Ishibashi K (2022) Tomato brown rugose fruit virus resistance generated by quadruple knockout of homologs of *TOBAMOVIRUS MULTIPLICATION1* in tomato. Plant Physiol 189:679–686
15. Fauser F, Schiml S, Puchta H (2014) Both CRISPR/Cas-based nucleases and nickases can be used efficiently for genome engineering in Arabidopsis thaliana. Plant J 79:348–359
16. Lei Y, Lu L, Liu H-Y, Li S, Xing F, Chen LL (2014) CRISPR-P: a web tool for synthetic single-guide RNA design of CRISPR-system in plants. Mol Plant 7:1494–1496
17. Sun HJ, Uchii S, Watanabe S, Ezura H (2006) A highly efficient transformation protocol for micro-tom, a model cultivar for tomato functional genomics. Plant Cell Physiol 47:426–431

Chapter 2

Silencing Host Factors with Hairpin RNA for Virus Resistance

Penghuan Rui, Jianping Chen, Fei Yan, and Guanwei Wu

Abstract

Plants rely on a multi-layered innate immune system to defend against abiotic and biotic, with RNA interference (RNAi) serving as a core antiviral mechanism. RNAi generates small interfering RNAs (siRNAs) that direct sequence-specific degradation and translational inhibition of viral RNAs. However, conventional strategies involving insertion of viral genes into the plant genome pose potential biosafety concerns. By contrast, RNAi-mediated silencing of endogenous host susceptibility genes offers distinct advantages, especially when combined with natural recessive resistance alleles. This chapter describes a detailed protocol for hairpin RNA (hpRNA) construct design and application, providing a practical reference for researchers screening and identifying host factors required for plant virus infection.

Key words RNA interference, Hairpin RNA, Plant virus

1 Introduction

Plant virus infection proceeds through sequential steps: entry into host cells, genome uncoating, viral protein translation, nucleic acid replication, progeny virion assembly, intercellular movement, systemic spread, and transmission to new hosts. To complete these processes, plant viruses rely on host susceptibility factors encoded by susceptibility genes (S-genes) [1]. Disruption of these essential functions confers virus resistance. Because the wild-type susceptible allele is typically dominant, loss-of-function mutations in S-genes give rise to recessive resistance [2], which can also arise from mutations in negative regulators of plant defense [3]. Such mutations interrupt critical virus-host interactions, blocking infection. Investigating recessive resistance thus provides valuable insights into viral pathogenesis and plant defense mechanisms [4].

In plants, RNA interference functions as a core antiviral mechanism. Small interfering RNAs of 21–22 nucleotides (nt) mediate sequence-specific degradation of target RNAs [5], while longer

Guanwei Wu and Xiaofei Cheng (eds.), *Plant Antiviral Resistance: Methods and Protocols*, Methods in Molecular Biology, vol. 3039, https://doi.org/10.1007/978-1-0716-5300-5_2,

siRNAs (24–26 nt) facilitate systemic signaling of post-transcriptional gene silencing (PTGS) and other silencing pathways [6–8]. Expression of self-complementary hairpin RNA (hpRNA) in transgenic plants represents a particularly PTGS strategy [9, 10]. hpRNA consists of two inverted repeat sequences that form a double-stranded stem, separated by a single-stranded loop. Incorporating a spliceosomal intron within this loop to generate intron-containing hpRNA (ihpRNA) can boost silencing efficiency to approximately 90% [10].

We applied this approach to construct an ihpRNA targeting the susceptibility gene, *feronia* from *Nicotiana benthamiana* (*NbFER*) [11]. Turnip mosaic virus, with its broad host range, serves as an ideal pathosystem for validating resistance [12, 13]. This chapter provides a robust framework for leveraging endogenous *S*-genes to engineer virus resistance in plants.

2 Materials

2.1 Construction of Hairpin RNA Vector

1. PFGC-5941 vector (*see* **Note 1**).
2. Fast-digesting enzymes BamH I and Xbal I.
3. 2× PrimeSTAR GXL Premix.
4. Mastercycler nexus GSX1.
5. Electrophoresis apparatus.
6. Constant temperature water bath.
7. HiPure gel pure DNA mini kit.
8. Agarose gel imaging system GelRed.

2.2 Preparation of Recombinant Plasmids

1. Homologous recombination enzyme.
2. *Escherichia coli* strain DH5α.
3. Ultra-clean workbench.
4. Luria-Bertani (LB) medium: Dissolve 10 g tryptone, 5 g yeast extract, and 10 g NaCl in 1 L ddH_2O. Adjust pH to 7.0 with 3 M NaOH. For solid medium, add 20 g agar per liter. Autoclave at 115 °C for 20 min (*see* **Note 2**).
5. Constant temperature shaker.
6. Constant temperature incubator.
7. Plasmid extraction kit.
8. Antibiotic stocks: rifampicin (25 mg/mL in methanol), kanamycin (100 mg/mL).

2.3 Agroinfiltration of Agrobacterium

1. *Agrobacterium tumefaciens* strain GV3101.
2. Gene pulser X cell electroporation system.
3. Electroporation cuvette.
4. *Nicotiana benthamiana* plants.
5. Injection syringe (without needles).
6. Agrobacterium infiltration buffer: 10 mM $MgCl_2$, 10 mM MES (pH 5.6), 150 mM acetosyringone (*see* **Note 3**).
7. PCB301-TuMV-GFP plasmid [14] (*see* **Note 4**).
8. UV lamp.

2.4 Protein Extraction and SDS-Polyacrylamide Gel Electrophoresis

1. Protein extraction buffer: 0.01 M PBS, 2% Na_2SO_3, 0.05% NP-40, 0.1% β-mercaptoethanol (*see* **Note 5**).
2. Loading buffer: 300 mM Tris–HCl (pH 6.8), 0.1% bromophenol blue, 8% (w/v) SDS, 25% β-mercaptoethanol, 40% (v/v) glycerol.
3. Resolving gel: 1.5 M Tris–HCl, pH 8.8, 10% (w/v) SDS, 30% acrylamide/bis solution (29:1), 10% ammonium persulfate solution (APS), N, N, N, N′-tetramethyl-ethylenediamine (TEMED).
4. Stacking gel: 0.5 M Tris–HCl, pH 6.8, 10% (w/v) SDS, 30% acrylamide/bis solution (29:1), 10% APS, TEMED.
5. SDS-PAGE running buffer: dissolve 1 g SDS, 14.4 g glycine and 3.03 g tris base in 900 mL ddH_2O, and then set the volume to 1 L.
6. Pre-stained protein marker, 14–120 kDa.

2.5 Western Blot

1. Tris-buffered saline with Tween (TBS-T) (1×): 0.25 M NaCl, 3 mM Tris–HCl, pH 7.5, 1% (v/v) Tween 20.
2. Polyvinylidine fluoride membrane.
3. Blocking buffer: 5% (w/v) non-fat milk powder in 1× TBS-T.
4. Primary antibodies: anti-TuMV CP antibody, anti-myc antibody.
5. Secondary antibodies: anti-mouse/rabbit IgG conjugated to HRP.
6. Developing reagent: HRP conjugate substrate kit.
7. Trans-Blot eBlot™ L1.
8. Amersham imager 680.
9. Coomassie brilliant blue staining solution (1 L): 2.5 g R250, 450 mL methyl alcohol, 100 mL glacial acetic acid, 450 mL ddH_2O.
10. Coomassie brilliant blue decolorization solution (1 L): 450 mL methyl alcohol, 450 mL ddH_2O, 100 mL glacial acetic acid.

11. Membrane transfer buffer: 25 mM Tris, 190 mM glycine, 20% methanol. For proteins larger than 80 kD, we recommend that SDS be included at a final concentration of 0.1%.

2.6 Real-Time Quantitative PCR

1. RNA extraction kit.
2. Tissue lyser.
3. Reverse transcription kit.
4. Applied biosystems real-time PCR instruments.
5. SYBR qPCR master mix.

3 Methods

3.1 Construction of Hairpin RNA Vector

1. Insert the 1353 bp chalcone synthase intron from Petunia hybrida into PFGC-5941. Clone the 221 bp NbFER fragment (Niben101Scf07619g00006.1, 1–221 nt CDS) downstream of the intron in antisense orientation using BamHI sites, and upstream in sense orientation using the XbaI site, with the following primers:
   BamHI-NbFER-F: CATTTGGAGAGGACACGCGG GATCCCGCAGCATCAGAGATTGATG;
   BamHI-NbFER-R: ATTGGGCGCGCCCCATGCGG GATCCCGATGACAGAAGGCAGTAAA;
   XbaI-NbFER-F: CAATTTGCAGGTATTTGGCTCTA GAGCATGACAGAAGGCAGTAAA;
   XbaI-NbFER-R: GGTCTTAATTAACTCTCTAGAGCAG CATCAGAGATTGATG. Generate hpGUS control construct using the same strategy. *See* Fig. 1 for diagram.
2. Linearize PFGC-5941 vector: Mix 5 μg PFGC-5941, 5 μL *Bam*H I Fast-restriction enzyme, 5 μL corresponding buffer, and ddH_2O to 50 μL. Incubate at 37 °C for 30 min (*see* **Note 6**).
3. Amplify BamH I-NbFER fragment: Use *N. benthamiana* cDNA as template. Prepare 50 μL PCR with: 25 μL 2× PrimeSTAR GXL Premix, 2.5 μL each of 10 μM forward/reverse primers (BamHI-NbFER-F/R), 1 μL cDNA, and 19 μL ddH_2O. Cycle: 98 °C 3 min; 35× (98 °C 30 s, 55 °C 30 s, 68 °C 1 min/kb); 68 °C 10 min.

Fig. 1 Chimeric hairpin RNA (hpRNA)

4. Purify the target fragments: Separate digested vector and PCR products by agarose gel electrophoresis. Excise target bands; purify via gel extraction kit. Determine concentration.
5. Perform homologous recombination: Combine 1 μL recombination enzyme, 2 μL recombination buffer, 2 μL linearized vector, and 5 μL purified PCR product. Incubate at 37 °C for 30 min.
6. Transform *E. coli* DH5α and screen: Add recombination product to competent cells; incubate on ice 30 min (*see* **Note 7**). Heat shock at 42 °C for 50 s; cool on ice 2 min. Add 1 mL LB, resuspend, and incubate at 37 °C for 30 min (*see* **Note 8**). Plate on kanamycin-LB medium, sequence-verify positive clones, and extract plasmid.
7. Digest recombinant plasmid with XbaI I. Re-amplify XbaI-NbFER fragment using same PCR system. Repeat recombination, transformation, and sequencing to obtain hpNbFER vector.

3.2 Transformation of Agrobacterium by Electroporation

1. Add 2 μL of the hpNbFER recombinant plasmid to *Agrobacterium* GV3101 competent cells that have been melted on ice.
2. Set the electroporation instrument to the "Agrobacterium" mode. Transfer the *Agrobacterium*-plasmid mixture to an electroporation cuvette and run the instrument (*see* **Note 9**).
3. After allowing it to stand at room temperature for 2 min, add 700 μL of non-resistant LB culture medium, mix thoroughly, then transfer the mixture to a 2 mL centrifuge tube and incubate it in a 28 °C constant temperature shaker for 2–3 h.
4. Centrifuge the collected transformed cells and then spread them on a solid LB medium containing 100 μg/mL kanamycin, and 25 μg/mL rifampicin, and incubate for 2 days at 28 °C.
5. Select individual colonies for PCR verification and then subculture and preserve the positive clones.

3.3 Agrobacterium Infiltration of N. benthamiana Leaves

1. Collect 2 mL of subcultured positive *Agrobacterium* cells by centrifugation at 5000 rpm for 5 min. Wash the cells twice with Agrobacterium infiltration buffer, then resuspend in 2 mL agrobacterium infiltration buffer. Determine the bacterial concentration using a spectrophotometer (*see* **Note 10**).
2. Adjust the OD_{600} of each Agrobacterium culture to 0.4 using agrobacterium infiltration buffer.
3. Specifically, adjust the OD_{600} of *Agrobacterium* carrying hpGUS or hpNbFER to 0.4, and that of *Agrobacterium* carrying PCB301-TuMV-GFP to 0.2 (*see* **Note 11**).

4. Mix *Agrobacterium* containing PCB301-TuMV-GFP plasmid with *Agrobacterium* containing hpNbFER or hpGUS recombinant plasmid at a 1:1 ratio.
5. Infiltrate middle leaves of 4- to 6-week-old *N. benthamiana* plants using a 1-mL syringe without the needle (*see* **Note 12**).
6. Visualize GFP derived from TuMV-GFP infection under UV lamp and photograph at 3 days post-infiltration. Harvest leaf discs with a 2-cm diameter hole puncher for protein and RNA extraction.

3.4 Protein Extraction and SDS-Polyacrylamide Gel Electrophoresis

1. Grind liquid nitrogen-frozen leaf samples into a powder. Add 100 μL protein extraction buffer and 25 μL loading buffer; vortex to mix. Incubate in a 95 °C water bath for 10 min to denature proteins (*see* **Note 13**).
2. Cool the denatured protein extract on ice, then centrifuge at 12,000 rpm for 5 min at 4 °C. Transfer the supernatant to a 2 mL microcentrifuge tube.
3. Mount the gel plate onto the electrophoresis tank, add 1 L SDS-PAGE running buffer, and carefully remove the comb. Load 10 μL of the protein supernatant into each well. Run at 80 V constant voltage until dye front reaches gel bottom (~2 h).

3.5 Western Blot

1. Cut a polyvinylidene fluoride (PVDF) membrane (8.5 × 4.5 cm, *see* **Note 14**) and activate in methanol for 2 min.
2. After electrophoresis, remove gel and assemble blotting sandwich per manufacturer instructions (Trans-Blot eBlot™ L1).
3. Insert cassette into Trans-Blot module (*see* **Note 15**) and start transfer program.
4. After transfer, block PVDF membrane in blocking buffer for 30 min at room temperature.
5. Dilute primary antibodies (anti-TuMV CP rabbit, anti-myc mouse) to 1:5000 in 20 mL blocking buffer. Incubate membrane overnight at 4 °C (*see* **Note 16**).
6. Wash membrane four times with 1× TBS-T, 5 min each (*see* **Note 17**).
7. Dilute HRP-conjugated secondary antibody (anti-rabbit/mouse IgG) to 1:10,000 in 20 mL blocking buffer. Incubate membrane for 1 h at room temperature.
8. Wash membrane five to eight times with 1× TBS-T, 5 min each.
9. Mix 100 μL each of HRP Conjugate Substrate A and B solutions; apply evenly to membrane. Image with Amersham Imager 680 (refer to system manual).
10. After imaging, stain membrane with Coomassie brilliant blue for 5 min (*see* **Note 18**), then destain for 10 min.

3.6 Quantitative Reverse Transcription PCR (qRT-PCR)

1. Freeze leaf samples in liquid nitrogen, grind with tissue lyser, keep on ice, add lysis buffer from RNA extraction kit.
2. Extract total RNA using TransGen Biotech RNA Extraction Kit (follow manual).
3. Quantify RNA, reverse transcribe 1 μg total RNA using TransGen Biotech Reverse Transcription Kit to synthesize cDNA.
4. Re-quantify cDNA and dilute to 100 ng/μL (*see* **Note 19**).
5. Set up qPCR: combine 6 μL cDNA, 1 μL forward primer, 1 μL reverse primer, 10 μL ddH_2O, and 18 μL SYBR qPCR Master Mix; mix and aliquot 10 μL into three tubes for technical replicates. Run program per instrument/reagent instructions (*see* **Note 20**).
6. Analyze qPCR data using $\Delta\Delta C_t$ method; perform Student's t-test for statistical analysis.

3.7 Expected Results

The *NbFER* gene is a membrane-bound receptor critical for plant growth, development and stress resistance [11]. Constructs expressing either hpGUS (control) or hpNbFER are co-infiltrated with TuMV-GFP into discrete leaf regions. At 3 days post-agroinfiltration (dpai), hpNbFER-treated regions display attenuated GFP fluorescence compared to hpGUS controls (Fig. 2a). NbFER transcript levels are reduced to ~70% of control by hpNbFER silencing (Fig. 2b). Western blot and RT-qPCR analyses further demonstrate significantly decreased TuMV CP accumulation in NbFER-silenced tissue (Fig. 2c, d).

4 Notes

1. PFGC-5941 vector plasmid is kept in our laboratory.
2. Avoid excessive sterilization temperature for LB medium: Autoclaving at 115 °C for 20 min is sufficient. Higher temperatures will cause medium deterioration.
3. Agrobacterium infiltration medium should be prepared fresh before use. Store stock solutions for the medium in a cool, dark place.
4. The PCB301-TuMV-GFP plasmid (GenBank Accession No.: NC_002509) is preserved in our laboratory.
5. Unless otherwise specified, all reagents and buffers are prepared with sterile ultrapure water at room temperature. Take protective measures when preparing reagents. Store the protein extraction buffer at 4 °C.
6. Restriction enzyme efficiency varies by brand; adjust the digestion time according to the product manual. Extending the digestion time appropriately improves linearization efficiency.

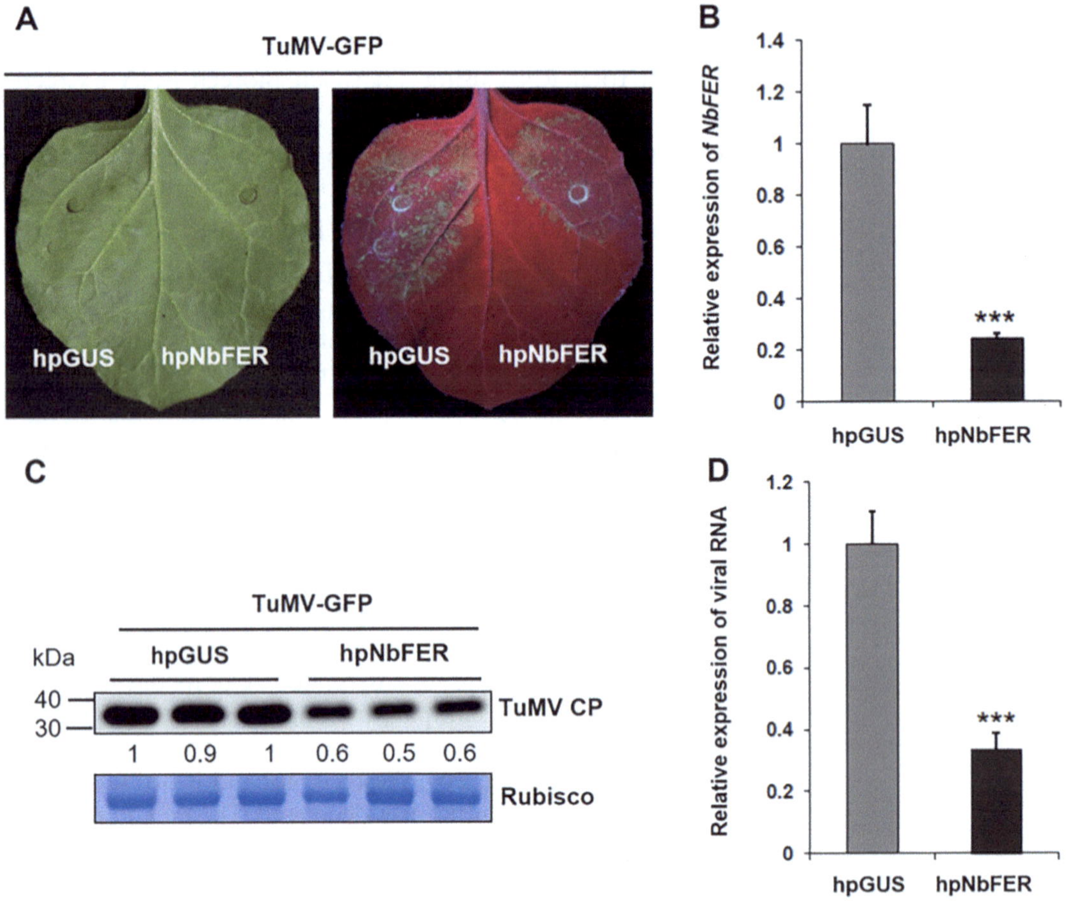

Fig. 2 Silencing NbFER impairs TuMV infection. (**a**) GFP fluorescence in *N. benthamiana* leaves co-expressing TuMV-GFP with hpGUS or hpNbFER at 3 days post-agroinfiltration (dpai). (**b**) *NbFER* transcript levels in samples from (**a**) at 3 dpai, measured by RT-qPCR. (**c**) Western blot analysis of TuMV CP and myc-tagged proteins in samples from (**a**) at 3 dpai. Relative CP band intensities (quantified by ImageJ) were normalized to Rubisco large subunit (CBB R-250 stain). (**d**) TuMV RNA accumulation in samples from (**a**) at 3 dpai, measured by RT-qPCR. Data are mean $\pm$ SD ($n = 5$ independent plants). Statistical significance was determined by two-tailed paired Student's t-test (***$P < 0.001$)

7. Thaw *E. coli* competent cells on ice; rapid temperature changes will kill competent cells and reduce transformation efficiency.
8. Incubate transformed *E. coli* into a 37 °C shaker for no more than 30 min. Prolonged incubation leads to excessive bacterial replication and increases false-positive clones.
9. Do not pre-cool electroporation cuvettes on ice before use. Condensation on the cuvette wall (caused by temperature differences) will cause short circuits and transformation failure.
10. When collecting *Agrobacterium* cells, keep the centrifugation speed ≤1000× *g*. Higher centrifugal force reduces bacterial viability or causes cell death.

11. The OD_{600} of *Agrobacterium* carrying PCB301-TuMV-GFP should not exceed 0.1; excessive viral replication will mask the effect of the target gene on viral infection.
12. To reduce experimental errors, infiltrate the control (hpGUS) and experimental (hpNbFER) bacterial suspensions into the left and right halves of the same *N. benthamiana* leaf (*see* Fig. 2). Complete infiltration of the entire leaf half is unnecessary. Avoid cross-contamination between different bacterial suspensions; wipe excess bacterial liquid from the leaf surface after infiltration.
13. Immediately place tissue lyser-ground samples on ice and add lysis buffer promptly to prevent moisture absorption and sample degradation.
14. Activate the PVDF membrane with methanol before transferring, then rinse with transfer buffer before use.
15. The Trans-Blot eBlot™ L1 is a rapid transfer system from GeneScript (15 min transfer time). Use the dedicated transfer buffer for this system; do not mix with wet-transfer buffers, as this causes experimental failure or instrument damage. Refer to the instrument manual for operation.
16. Incubate primary antibodies at 4 °C overnight (static incubation) to reduce non-specific antibody binding and minimize background noise during imaging.
17. Store Tris-Buffered Saline (TBS-T) in the dark (due to Tween 20). Use TBS-T immediately after preparation; degraded Tween 20 reduces membrane washing efficiency and increases imaging background.
18. Stain the PVDF membrane with Coomassie brilliant blue for 5–10 min; prolonged staining increases background. Perform staining in a fume hood, as the staining solution contains volatile toxic reagents.
19. Diluting cDNA to 100 ng/μL is critical: High initial cDNA concentration reduces result significance or eliminates statistical differences.
20. Verify the amplification efficiency and specificity of qPCR primers before use to ensure reliable results.

Acknowledgment

This research was funded by the grants from the National Natural Science Foundation of China (3207016), the China Agriculture Research System of MOF and MARA (CARS-24-C-04), and the K. C. Wong Magna Fund in Ningbo University.

References

1. Mäkinen K (2019) Plant susceptibility genes as a source for potyvirus resistance. Ann Appl Biol 176:122–129
2. de Ronde D, Butterbach P, Kormelink R (2014) Dominant resistance against plant viruses. Front Plant Sci 5:307
3. Truniger V, Aranda MA (2009) Recessive resistance to plant viruses. Adv Virus Res 75:119–159
4. Kang BC, Yeam I, Jahn MM (2005) Genetics of plant virus resistance. Annu Rev Phytopathol 43:581–621
5. Palauqui JC, Elmayan T, Pollien JM et al (1997) Systemic acquired silencing: transgene-specific post-transcriptional silencing is transmitted by grafting from silenced stocks to non-silenced scions. EMBO J 16: 4738–4745
6. Hamilton A, Voinnet O, Chappell L et al (2015) Two classes of short interfering RNA in RNA silencing. EMBO J 21:4671–4679
7. Hall IM, Noma K, Grewal SI (2003) RNA interference machinery regulates chromosome dynamics during mitosis and meiosis in fission yeast. Proc Natl Acad Sci USA 100:193–198
8. Zilberman D, Cao X, Jacobsen SE (2003) Argonaute4 control of locus-specific siRNA accumulation and DNA and methylation. Science 299:716–719
9. Smith NA, Singh SP, Wang MB et al (2000) Total silencing by intron-spliced hairpin RNAs. Nature 407:319–320
10. Wesley SV, Helliwell CA, Smith NA et al (2001) Construct design for efficient, effective and high-throughput gene silencing in plants. Plant J 6:581–590
11. Rui P, Jia Z, Fang X et al (2025) A plant viral effector subverts FER-RALF1 module-mediated intracellular immunity. Plant Biotechnol J 23:2734–2751
12. Wu G, Fang X, Yu T et al (2024) Turnip mosaic virus pathogenesis and host resistance mechanisms in *Brassica*. Hortic Plant J 10:947–960
13. Wu G, Jia Z, Ding K et al (2022) Turnip mosaic virus co-opts the vacuolar sorting receptor VSR4 to promote viral genome replication in plants by targeting viral replication vesicles to the endosome. PLoS Pathog 18: e1010257
14. Wu G, Cui X, Chen H et al (2018) Dynamin-like proteins of endocytosis in plants are coopted by potyviruses to enhance virus infection. J Virol 92:e01320–18

Chapter 3

Heritable Tissue-Culture-Free Gene Editing in *Nicotiana benthamiana* Using a Meristem-Invading Virus Vector

Tetsuya Yoshida and Kazuhiro Ishibashi

Abstract

Gene editing can be achieved using sequence-specific nucleases. This protocol describes a plant gene editing method that eliminates the need for tissue culture by employing a virus-based delivery system. Tobacco ringspot virus (TRSV) can access meristematic tissues in infected plants, enabling the introduction of site-directed mutations into germline cells. This allows for heritable gene modification in the model plant *Nicotiana benthamiana* without tissue culture.

Key words Gene editing, Virus vector, SpCas9, Tobacco ringspot virus, Tissue-culture-free

1 Introduction

Plant gene editing typically involves the introduction or expression of sequence-specific nucleases, such as CRISPR-Cas9 components. This is commonly achieved through tissue culture-mediated transformation. However, tissue culture is labor-intensive and applicable to only a limited range of plant species or cultivars, making it a major bottleneck in the implementation of gene editing technologies in plants [1].

Virus vectors serve as valuable tools for the expression of foreign genes *in planta*, with their utility in gene editing having been recently investigated [2, 3]. Nevertheless, various technical obstacles—such as the considerable size of site-specific nuclease (SSN)-encoding genes (for instance, the *Streptococcus pyogenes Cas9* gene, *SpCas9*, which is approximately 4.1 kb long) and the inherent antiviral mechanisms present in plant shoot apical meristems [4]—have constrained the effectiveness of fully virus vector-mediated, tissue-culture-free gene editing approaches. The resistance of meristematic tissues to viral penetration can be partially mitigated by introducing multiple virus vectors encoding sgRNAs with functional motifs designed to enhance RNA mobility in plants

Guanwei Wu and Xiaofei Cheng (eds.), *Plant Antiviral Resistance: Methods and Protocols*, Methods in Molecular Biology, vol. 3039,
https://doi.org/10.1007/978-1-0716-5300-5_3,

into Cas9-expressing transgenic lines [5–13]. Conversely, although there have been limited successful instances of direct Cas9 expression from plant virus vectors within somatic cells [14–18], achieving such expression in germline cells remains a significant challenge.

In this section, we introduce a virus vector capable of both invading plant meristems and expressing Cas9 and sgRNA, thereby enabling heritable gene editing in non-transgenic *Nicotiana benthamiana* plants [19]. This virus vector, namely tobacco ring spot virus (TRSV) vector, has the potential to eliminate the need for tissue culture in plant gene editing applications (*see* **Note 1**).

2 Materials

1. TRSV vector plasmids (pTRSV1, pTRSV2-SpCas9) [19].
2. ALSV vector plasmids (pALSV1, pALSV2-NbRDR6) [19].
3. TRV vector plasmids (pTRV1, pTRV2-sgRNA) [5].
4. p19 RNA silencing suppressor plasmid (pRI-p19 or equivalent) [20].
5. Restriction enzymes such as MluI and SalI.
6. NEBuilder® HiFi DNA Assembly Master Mix.
7. Spectinomycin or kanamycin for bacterial selection.
8. LB medium.
9. Shaking incubator (37 °C and 28 °C).
10. *E. coli* for standard transformation.
11. Plasmid DNA isolation kit/reagents.
12. *Agrobacterium* strain C58C1.
13. Spectrophotometer (OD600 measurement).
14. Growth chamber for cultivating *Nicotiana benthamiana* plants under a 16 h light/8 h dark cycle at 25 °C.
15. 10 mM Tris–HCl pH 8.0.
16. 5 M NaCl.
17. 1 M MES-KOH pH 5.8.
18. 1 M $MgCl_2$.
19. 100 mM acetosyringone stock solution prepared in dimethyl sulfoxide.
20. Needleless syringe.
21. Synthesized oligonucleotides for sgRNA cloning.
22. DNA extraction kit/reagents from plants.

3 Methods

3.1 Plasmid Construction

1. For expression of TRSV RNA1 and the RNA silencing suppressor p19, pTRSV1 and pRI-p19, respectively, are readily available for any target genes. Gene-specific sgRNA sequence in pTRSV2-SpCas9 should be modified as described later (Fig. 1, *see* **Note 2**).
2. Digest pTRSV2-SpCas9 with MluI and SalI.
3. Mix 2 μL each of 10 μM TRSV2-GuideOligo containing 20-nt sequence targeting gene-of-interest and TRSV2-sgOligoRv, 1 μL of 5 M NaCl, and 95 μL of 10 mM Tris–HCl (pH 8.0). Boil for 3 min at 100 °C, then allow to cool gradually.
4. Mix 150 ng of MluI- and SalI-digested pTRSV2-SpCas9, 5 μL of the annealed oligo DNA, and water to 10 μL, and then with 10 μL of NEBuilder® HiFi DNA Assembly Master Mix. Incubate at 50 °C for 60 min. Transform 2 μL of the mixture into chemically competent *Escherichia coli* cells. Select spectinomycin-resistant colonies and isolate plasmid DNA.
5. Transform *Agrobacterium* strain C58C1 with sequence-confirmed, purified plasmids and select spectinomycin-resistant colonies.

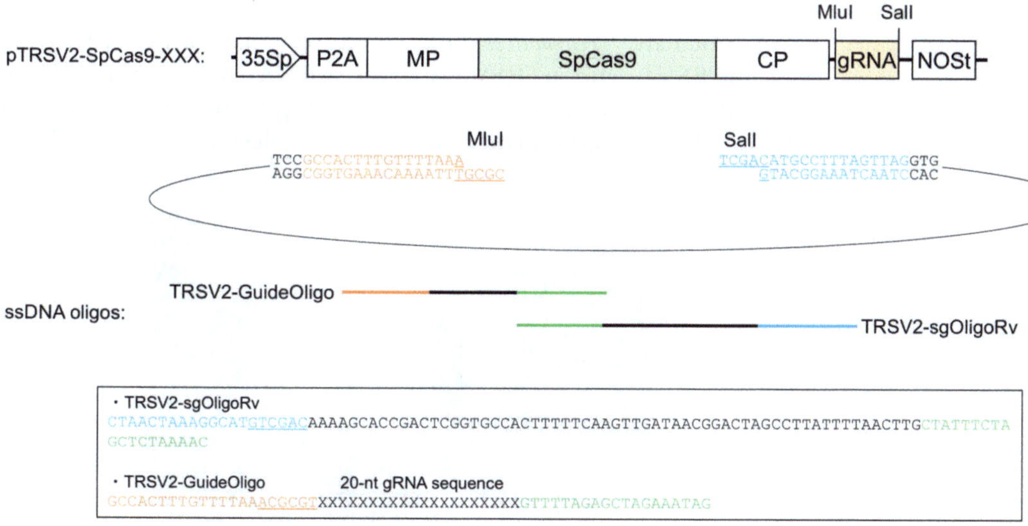

Fig. 1 Single guide RNA cloning into pTRSV2-SpCas9 vector
Synthesized oligo DNA for a target gene (TRSV2-GuideOligo with any 20-nt sequence) and the fixed-sequence oligo DNA TRSV2-sgOligoRv are cloned into MluI- and SalI-digested pTRSV2-SpCas9

3.2 Agroinoculation of TRSV

1. Grow *N. benthamiana* plants in a growth chamber under a 16-h light/8-h dark cycle at 25 °C.
2. Optional: To enhance the gene-editing efficiency, downregulation of the *NbRDR6* gene which contributes to the antiviral RNA silencing is recommended. For knockdown of *NbRDR6* expression, inoculate *Agrobacterium* mixtures harboring pRI-p19, pALSV1, and pALSV2-NbRDR6 into abaxial side of the first true leaves of 2-week-old *N. benthamiana* plants.
3. Inoculate *Agrobacterium* strains harboring pRI-p19, pTRSV1, or pTRSV2-SpCas9-sgRNA into 2 mL of LB medium containing either 50 μg/mL kanamycin or spectinomycin, as appropriate. Incubate at 28 °C with shaking at 200–220 rpm overnight (~16 h).
4. Collect the bacterial cells by centrifugation at 10,000× *g* for 1 min. Discard the supernatant and resuspend the pellet in 1 mL of agroinfiltration buffer (10 mM MES-KOH pH 5.8, 10 mM $MgCl_2$, 100 μM acetosyringone). Incubate the suspension at room temperature for 3–16 h in the dark to activate *vir* genes. Measure the optical density at 600 nm (OD_{600}) prior to agroinfiltration. The same procedure should be followed when preparing *Agrobacterium* cultures for ALSV or TRV inoculation.
5. Mix *Agrobacterium* solutions for OD_{600} of 0.2 and 0.5 each for pRI-p19 and viral infectious clones, respectively, before inoculation.
6. Inoculate *Agrobacterium* mixtures harboring pRI-p19, pTRSV1, and pTRSV2-SpCas9-sgRNA into abaxial side of the fifth or sixth true leaves of approximately 4-week-old *N. benthamiana* plants by 1-mL needleless syringe (Fig. 2).

Fig. 2 Agroinfiltration into *N. benthamiana* leaves

7. Co-inoculating a TRV vector expressing sgRNA with a TRSV vector carrying SpCas9 can improve gene editing efficiency, especially for sgRNAs that do not efficiently work with TRSV vector. Prepare *Agrobacterium* strains with pTRV1, pTRV2-sgRNA, and mix them with those containing TRSV vectors for co-inoculation.

3.3 Induction and Selection of Lateral Shoots with High Level Gene Editing

1. Grow inoculated plants for about 3 weeks (*see* **Note 3**). When gene editing is induced in shoot apical meristem tissues, the 10th upper leaves from the TRSV-inoculated leaves—where the infecting TRSV vector typically lacks the SpCas9 sequence insert—often carry the introduced mutations. If mutation frequencies are low, the likelihood of obtaining gene-edited progeny is less than 1%. Growing and screening lateral shoots can facilitate the isolation of highly gene-edited tissues. To do this, cut the primary shoots of inoculated plants above the 10th upper leaves.
2. Harvest a leaf or punch a small section from each lateral shoot for DNA extraction. Examine the extent of gene editing by high-throughput sequencing, restriction enzyme digestion, hybridization-based mutation detection, electric mobility and so on. Select highly gene-edited lateral shoots and cut off others for facilitating the growth of selected shoots when necessary.
3. Collect seeds from selected shoots. TRSV-infected *N. benthamiana* plants often have low fertility, especially when *NbRDR6* expression was knocked down. If highly edited shoots don't produce seeds, cross-pollinate with pollen from healthy plants. F1 plants are generally fertile, and homozygous gene-edited individuals can be obtained in the F2 generation.
4. Sow the seeds and analyze the genotype. Though at a low frequency, TRSV may be seed-transmitted. Transmitted TRSV can be detected by RNA-based methods such as RT-PCR or protein-based methods such as ELISA.

4 Notes

1. TRSV exhibits a comparatively broad host range. While host plants of TRSV present potential candidates for this approach, successful application has thus far been limited to *N. benthamiana*.
2. The sgRNA sequence plays a key role in gene editing when using the TRSV vector, as compared to conventional transformation-mediated expression of SpCas9. Testing multiple sgRNA sequences for each target gene is recommended.

3. It is important to monitor plants for overall vigor throughout this process, as both TRSV and TRV vectors can induce varying degrees of physiological stress that may impact development and seed set. Careful management of growth conditions—such as optimal light intensity and temperature—can improve plant recovery and fertility. In some cases, supplementing additional nutrients or adjusting light intensities may further support healthy lateral shoot development and subsequent seed production.

Acknowledgment

This work was supported in part by the Cross-ministerial Strategic Innovation Promotion Program (funding agency: Bio-oriented Technology Research Advancement Institution), "Technologies for Smart Bio-industry and Agriculture" and "Building a Resilient and Nourishing Food Chain for a Sustainable Future" (grant number JPJ012287).

Competing Interest Statement T.Y. and K.I. are named as inventors on a patent application related to this work.

References

1. Altpeter F, Springer NM, Bartley LE et al (2016) Advancing crop transformation in the era of genome editing. Plant Cell 28:1510–1520
2. Uranga M, Daròs JA (2023) Tools and targets: the dual role of plant viruses in CRISPR–Cas genome editing. Plant Genome 16:e20220
3. Zhang C, Liu S, Li X, Zhang R, Li J (2022) Virus-induced gene editing and its applications in plants. Int J Mol Sci 23:10202
4. Bradamante G, Mittelsten Scheid O, Incarbone M (2021) Under siege: virus control in plant meristems and progeny. Plant Cell 33:2523–2537
5. Ellison EE, Nagalakshmi U, Gamo ME, Huang P, Dinesh-Kumar S, Voytas DF (2020) Multiplexed heritable gene editing using RNA viruses and mobile single guide RNAs. Nat Plants 6:620–624
6. Lei J, Li Y, Dai P, Liu C, Zhao Y, You Y, Qu Y, Chen Q, Liu X (2022) Efficient virus-mediated genome editing in cotton using the CRISPR/Cas9 system. Front Plant Sci 13:1032799
7. Li T, Hu J, Sun Y et al (2021) Highly efficient heritable genome editing in wheat using an RNA virus and bypassing tissue culture. Mol Plant 14:1787–1798
8. Uranga M, Aragonés V, Selma S, Vázquez-Vilar M, Orzáez D, Daròs JA (2021) Efficient Cas9 multiplex editing using unspaced sgRNA arrays engineering in a *Potato virus X* vector. Plant J 106:555–565
9. Nagalakshmi U, Meier N, Liu JY, Voytas DF, Dinesh-Kumar SP (2022) High-efficiency multiplex biallelic heritable editing in Arabidopsis using an RNA virus. Plant Physiol 189:1241–1245
10. Oh Y, Nagalakshmi U, Dahlbeck D, Koehler N, Cho MJ, Dinesh-Kumar SP, Staskawicz BJ (2025) Heritable virus-induced germline editing in tomato. Plant J 122:e70115
11. Kang B, Lee S, Ko D, Venkatesh J, Kwon JK, Kim H, Kang BC (2025) Virus-induced systemic and heritable gene editing in pepper (*Capsicum annuum* L.). Plant J 122:e70257
12. Liu D, Ellison EE, Myers EA, Donahue LI, Xuan S, Swanson R, Qi S, Prichard LE, Starker CG, Voytas DF (2024) Heritable gene editing in tomato through viral delivery of isopentenyl transferase and single-guide RNAs to latent axillary meristematic cells. Proc Natl Acad Sci USA 121:e2406486121

13. Lee H, Baik JE, Kim KN (2025) Development of an efficient and heritable virus-induced genome editing system in *Solanum lycopersicum*. Hortic Res 12:uhae364
14. Ariga H, Toki S, Ishibashi K (2020) Potato virus X vector-mediated DNA-free genome editing in plants. Plant Cell Physiol 61:1946–1953
15. Ma X, Zhang X, Liu H, Li Z (2020) Highly efficient DNA-free plant genome editing using virally delivered CRISPR–Cas9. Nat Plants 6: 773–779
16. Liu Q, Zhao C, Sun K, Deng Y, Li Z (2023) Engineered biocontainable RNA virus vectors for non-transgenic genome editing across crop species and genotypes. Mol Plant 16:616–631
17. Zhao C, Lou H, Liu Q, Pei S, Liao Q, Li Z (2024) Efficient and transformation-free genome editing in pepper enabled by RNA virus-mediated delivery of CRISPR/Cas9. J Integr Plant Biol 66:2079–2082
18. Lee SY, Kang B, Venkatesh J, Lee JH, Lee S, Kim JM, Back S, Kwon JK, Kang BC (2024) Development of virus-induced genome editing methods in Solanaceous crops. Hortic Res 11: uhad233
19. Yoshida T, Ishikawa M, Toki S, Ishibashi K (2024) Heritable tissue-culture-free gene editing in *Nicotiana benthamiana* through viral delivery of SpCas9 and sgRNA. Plant Cell Physiol 65:1743–1750
20. Kaya H, Ishibashi K, Toki S (2017) A Split *Staphylococcus aureus* Cas9 as a compact genome-editing tool in plants. Plant Cell Physiol 58:643–649

Chapter 4

Pepper Veinal Mottle Virus (PVMV)-Based Vector for Expressing Pathogen-Derived Resistance in Pepper

Changhui Deng, Na Li, and Hongguang Cui

Abstract

Pathogen-derived resistance (PDR) is an effective approach in the management of plant viruses in agriculture. It is achieved via expressing a viral protein or genomic fragment in a host plant. Numerous genetically modified crops expressing viral coat protein (CP) show good performance in defending target viruses. Unfortunately, the cultivated pepper species are recalcitrant to genetic transformation and regeneration. As an alternative, an attenuated virus-based vector can be used for the delivery of desired viral gene or genomic fragment into host plant to trigger PDR. Here, we have described in detail a protocol using an attenuated strain of pepper veinal mottle virus (PVMV)-based vector for the delivery of target virus *CP* in pepper, thus triggering PDR.

Key words Pathogen-derived resistance, Pepper veinal mottle virus, Viral vector, Pepper viruses, Golden Gate cloning, Coat protein

1 Introduction

Pepper (*Capsicum* spp.) is globally a significant spice and vegetable crop. Unfortunately, its production is severely constrained by a multitude of viral diseases [1, 2]. Up to 164 viral pathogens, classified into 34 genera across 17 families, have been reported in peppers [2, 3]. Pathogen-derived resistance (PDR), a concept first proposed by Sanford and Johnston in 1985, is an efficient strategy for the control of plant viruses, which is achieved through engineering host plants to express a viral gene or genomic fragment (in the form of hairpin structure in most cases) to confer virus resistance [4, 5]. Some genetically modified crops expressing robust PDR have been approved for commercial plantation, such as *Cucurbita pepo* against cucumber mosaic virus, zucchini yellow mosaic virus, and watermelon mosaic virus, *Carica papaya* against papaya ringspot virus, and *Solanum tuberosum* against potato virus Y, and potato leafroll virus [4]. However, a stable genetic

Guanwei Wu and Xiaofei Cheng (eds.), *Plant Antiviral Resistance: Methods and Protocols*, Methods in Molecular Biology, vol. 3039, https://doi.org/10.1007/978-1-0716-5300-5_4,

transformation along with efficient regeneration for cultivated pepper species are still challenging [6], thus restricting the application of PDR in pepper. This limitation prompted us to explore an alternative strategy for the delivery of viral gene or genomic fragment, such as using an attenuated virus-based expression vector.

Plant virus-based vectors can deliver foreign sequences into host plants for expression, serving as a powerful biotechnology in pest management via expressing a sequence of plant pathogen and insect pest [7, 8]. RNA silencing (RNA interference, RNAi), a conserved antiviral mechanism in plants, is a principal mechanism underlying the resistance. Theoretically, when an attenuated virus strain, harboring a cistron from a heterologous virus such as *Coat Protein* (*CP*), infect host plants, the resistance against the target virus will be established. Along with the infection by the recombinant virus, the replicative intermediate—double-stranded RNA (dsRNA) is formed. The dsRNA precursor is recognized and processed by Dicer-like (DCL) enzymes into virus-derived small interfering RNAs (vsiRNAs), followed by the assembly into RNA-induced silencing complex (RISC) [9]. The inserted sequence-derived siRNAs would guide the cleavage and degradation of RNA genomes of target virus in a sequence-specific manner [4, 5]. In addition, protein-mediated mechanisms may also contribute to the resistance. In particular, when viral CP is preexisted *in planta*, it would compete for binding sites in the genomes of invading cognate virus, interfering with viral genome translation or de-coating [4, 5].

Pepper veinal mottle virus (PVMV) is a member of the genus *Potyvirus* in the family *Potyviridae*, with a wide range of Solanaceae plants as its natural hosts [3, 10]. PVMV have a positive-sense, single-stranded RNA genome (+ssRNA). The +ssRNA genome contains a long open reading frame (ORF), and another short ORF (PIPO) embedded in the P3-coding region (Fig. 1a) [11, 12]. After translation, two polyproteins, including a large one (P1-to-CP) and another small one (P1-to-P3N-PIPO), are produced, and immediately proteolytically processed by three virus-encoded proteases (P1, HCPro, and NIa-Pro) into a dozen of viral factors (Fig. 1a). Numerous documents show that potyviruses can tolerate foreign inserts at P1/HC-Pro and NIb/CP intercistronic junctions [13, 14]. This plasticity enables the virus to function as a vehicle for delivering foreign sequences into plant hosts for expression, thus bypassing the recalcitrance of stable genetic transformation and regeneration for some plant species.

Plant virus-based vectors are valuable biotechnological tools for the expression of foreign genes-of-interest (GOI) (Fig. 1b). Our group developed an infectious clone of PVMV isolate PVMV-HNu, which was used to confirm the etiology of foliar chlorosis and rugosity disease in pepper [15]. Then, we developed an attenuated strain of PVMV on a basis of a spontaneous site in 6K1

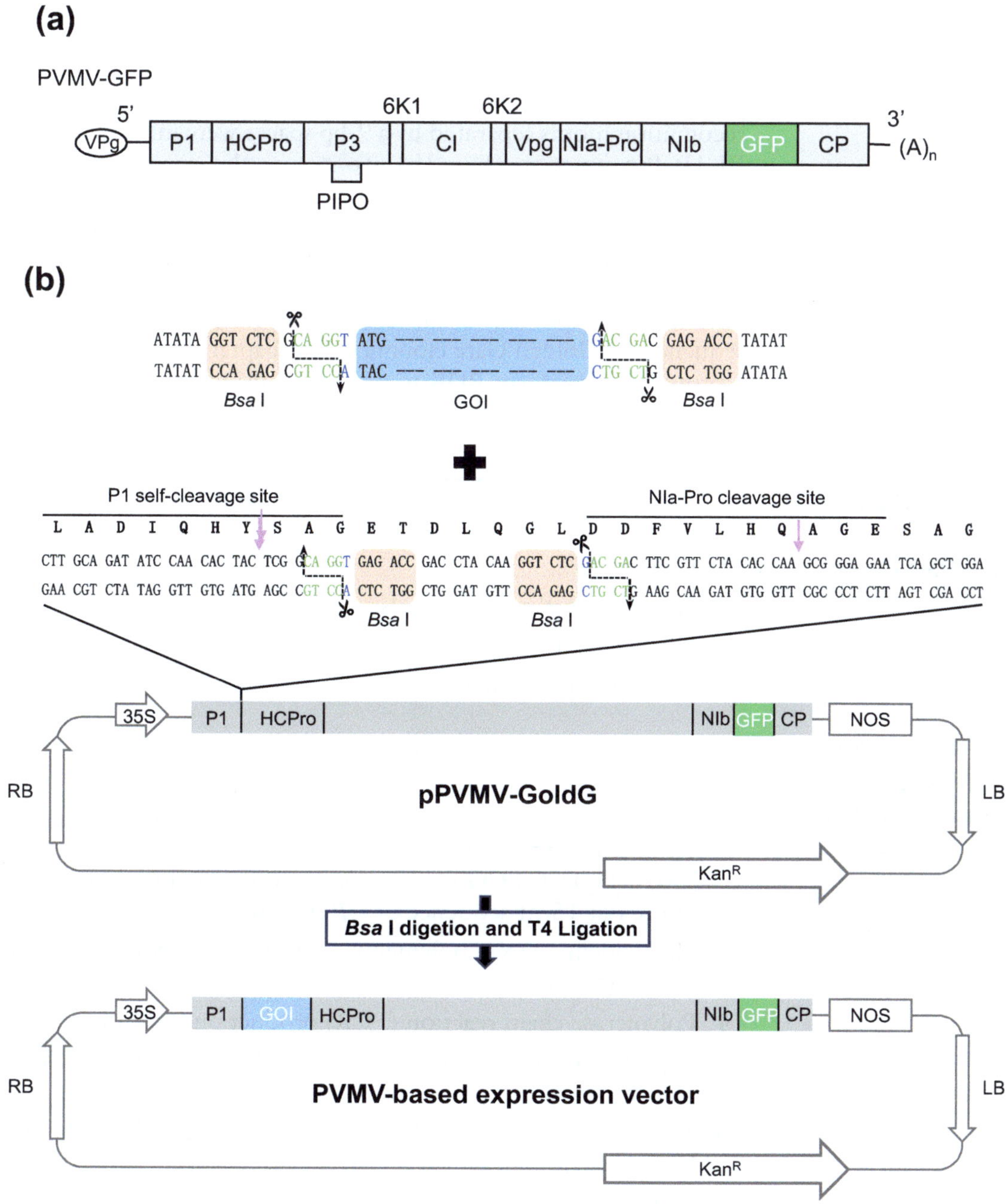

Fig. 1 Schematic of pepper veinal mottle virus (PVMV)-based vector having Golden Gate cloning feature. (**a**) Schematic of the genomic organization of GFP-tagged PVMV (PVMV-GFP). The two short horizontal lines indicate the 5′- and 3′-untranslated regions. The large box represents the long open reading frame (ORF); the smaller boxes represent viral factors which are produced through proteolytic cleavage of viral polyproteins by virus-encoded proteases. The green box represents a complete GFP-coding sequence, which is engineered into NIb-CP junction. Another short ORF (PIPO) embedded in the P3-coding region is shown. (**b**) Schematic diagrams showing pPVMV-GoldG and cloning of foreign GOI into the vector via Golden Gate cloning. P1's self-cleavage and NIa-Pro's trans-cleavage sites were indicated by purple arrows. The recognition motifs of *Bsa* I are shown in orange background. The cleavage sites by *Bsa* I are indicated by arrowheads. GOI gene-of-interest (foreign insert), 35S the cauliflower mosaic virus 35S promoter, Nos the nopaline synthase terminator, LB left border, RB right border

protein [16, 17]. Subsequently, we developed the attenuated strain-based expression vector (designated pPVMV-GoldG in this study), which is compatible with Golden Gate cloning. For pPVMV-GoldG, a tandem of *Bsa* I (type IIS restriction enzyme) recognition motifs separated by a 9-bp spacer were introduced into P1-HCPro junction (Fig. 1b). To ensure the proper release of heterologous protein from virus-encoded polyprotein, the original sequence encoding P1's self-cleavage peptide was retained, and another short sequence encoding NIa-Pro's cleavage peptide was incorporated into the upstream of HCPro (Fig. 1b) [16]. This optimized vector allows for convenient integration of foreign sequence via Golden Gate cloning. This chapter describes in detail a protocol to perform PDR in pepper by using the PVMV-based vector to express the *CP* cistron of target virus.

2 Materials

2.1 Plant Material

1. Pepper seeds.
2. *Nicotiana bethamiana* seeds.
3. Pots and soil.
4. Plant growth cabinet.

2.2 Construction of PVMV-Based Recombinant Vector

1. pPVMV-GoldG vector [16].
2. Primer design for amplifying the *CP* cistron of target virus, CP-BsaI-F: (5′ → 3′) ATATAGGTCTCG**CAGG**T + the first 15-nt sequence of *CP* cistron, CP-BsaI-R: (5′ → 3′) ATATAGGTCTCG**TCGT**C + reverse complementary sequence of the last 15-nt sequence of *CP* cistron.
3. High-fidelity DNA polymerase, buffer, and dNTP (Thermo Fisher Scientific).
4. Polymerase chain reaction (PCR) product purification kit.
5. Type IIS restriction enzyme *Bsa* I (20 U/μL), and 10× CutSmart buffer (NEB).
6. T4 DNA Ligase (5 U/μL), and 10× Ligation buffer (Thermo Fisher Scientific).
7. PCR cycler equipment.
8. Sterilized centrifuge tubes (1.5 mL, 10 mL).
9. Competent cells of *Escherichia coli* (DH10B strain).
10. Heating dry Bath.
11. Luria broth (LB) medium.
12. 100 mg/mL kanamycin stock solution.

13. Primer set for screening positive clones, PV-P1-F: (5′ → 3′) GCATACCTCATGGCTTGCTC, PV-HCPro-R: (5′ → 3′) AGAAGGGTTGTTCCTTCCTG.
14. Plasmid purification kit.

2.3 Agrobacterium Infiltration

1. Competent cells of *Agrobacterium tumefaciens* (GV3101 strain).
2. Electroporation apparatus (Bio-Rad).
3. 20 mg/mL rifampicin stock solution.
4. Microfuge.
5. Inoculation buffer: 10 mM $MgCl_2$, 10 mM MES (pH 5.7), 200 μM Acetosyringone.
6. Sterilized centrifuge tubes (2 mL).
7. 1 mL syringe.
8. UV lamp (LUYOR-3410, LUYOR).

2.4 Sap Rub-Inoculation

1. Mortar and pestle.
2. Absolute ethanol.
3. Corundum powder (240 mesh).
4. Inoculation buffer (working solution): 0.01 M KH_2PO_4, 0.01 M Na_2HPO_4.
5. Sterilized scissors and finger-cot.

3 Methods

3.1 Preparation of Pepper Seedlings

1. Store pepper seeds in a cold room (4 °C, darkness).
2. Soak seeds in sterilized double-distilled water for 1 day (*see* **Note 1**).
3. Spread pepper seeds evenly into a pot filled with soil, and then cover them with a thin layer of soil.
4. Maintain all pepper seedlings in a plant growth cabinet (16-h light, 25 °C and 8 h of darkness, 23 °C, 70% relative humidity) (*see* **Note 2**).

3.2 Construction of PVMV-Based Recombinant Vector Carrying the CP Cistron of Target Virus

1. Check whether the *CP* sequence of target virus contains *Bsa* I recognition motif (*see* **Note 3**). If it does, a synonymous mutation needs to be introduced to destroy the motif (*see* **Note 4**).
2. Amplify the complete *CP* sequence by primer set CP-BsaI-F/CP-BsaI-R (*see* **Note 5**). The resulting PCR product harboring *Bsa* I recognition motifs at both ends is purified with PCR product purification kit.

3. Set up a digestion–ligation reaction in a 1.5-mL tube (10 μL): purified PCR product (~60 ng), pPVMV-GoldG (~50 ng), T4 DNA ligase (5 U), *Bsa* I (16 U), 10× CutSmart buffer (1 μL), and 10× Ligation buffer (1 μL).
4. Incubate the digestion–ligation mix in a thermal cycler at 37 °C, 2–4 h for *Bsa* I digestion, followed by ligation reaction at 50 °C for 5 min.
5. Heat treatment inactivation for 10 min at 80 °C (*see* **Note 6**).
6. Add the obtained product (5 μL) into *E. coli* competent cells (30–40 μL). Place the mixture on ice (30 min), treat it for 45 s at 42 °C (heat shock), and incubate it on ice for 2 min.
7. Add 700 μL of LB medium (liquid, without antibiotics) to the mixture, and incubate it for 1 h in a shaker (37 °C, 180 rpm).
8. Spread the culture onto LB medium plate (solid, 100 μg/mL kanamycin), followed by inverted cultivation for 12–16 h at 37 °C (*see* **Note 7**).
9. Screen positive colonies by using primer set PV-P1-F/PV-HCPro-R. Inoculate a positive colony into 5 mL LB medium (liquid, 100 μg/mL kanamycin), followed by propagation overnight in a shaker (37 °C, 180 rpm) (*see* **Note 8**).
10. Extract plasmids with a plasmid purification kit, and determine plasmid concentration by using a spectrophotometer (*see* **Note 9**).
11. Plasmid verification by restriction digestion with *Mlu* I and *BamH* I, and Sanger sequencing (*see* **Note 10**).

3.3 Pre-inoculation with PVMV-Based Recombinant Vector

1. De-freeze competent cells of *Agrobacterium tumefaciens* (stored at −80 °C) on ice for 2–5 min (*see* **Note 11**).
2. Add 50 ng of plasmid DNA into 30–40 μL of competent cells, transfer the mix into an ice-cold electroporation cuvette, and incubate it on ice for 2 min.
3. Adjust the settings of electroporation apparatus for deliver a pulse of 2400 V, 200 Ω resistance, 25 μF capacitance.
4. Dry the outside of electroporation cuvette with a tissue, place it in the cuvette holder of electroporation apparatus, and deliver the electric pulse.
5. Add 700 μL of LB (without antibiotics) into the cuvette, transfer the cell suspension into a 1.5-mL centrifuge tube, and incubate it for 2 h at 28 °C, 180 rpm.
6. Spread the culture onto LB medium plate (100 μg/mL kanamycin, and 20 μg/mL rifampicin), followed by inverted cultivation for 36–48 h at 28 °C.
7. Screen positive colonies by using primer set PV-P1-F/PV-HCPro-R. Inoculate a positive colony into 5 mL LB medium (100 μg/mL kanamycin, 20 μg/mL rifampicin), and propagate it for 24–36 h in a shaker (28 °C, 180 rpm).

8. Transfer the culture into 2-mL centrifuge tube, followed by centrifuge at 5000 rpm for 5 min.
9. Discard the supernatant, and gently dissolve cell pellet in 1.5 mL inoculation buffer. Centrifuge at 5000 rpm for 5 min.
10. Repeat **step 9** once.
11. Remove the supernatant carefully, gently dissolve cell pellet in 1.5 mL inoculation buffer. Adjust OD_{600} of cell suspension to 1.0 (*see* **Note 12**).
12. Place cell suspension at room temperature for 1 h.
13. Infiltrate cell suspension into the newly expanded leaf of pepper seedlings at 1- or 2-leaf-stage by using 1-mL syringe without needle.
14. Monitor virus infection by green fluorescence observation with a UV lamp, as well as other molecular methods such as RT-PCR and Western blot.

3.4 Challenging Inoculation with Target Virus

1. Prepare agrobacterial culture harboring the clone of target virus, following the description in previous section. Infiltrate the newly expanded leaves of *Nicotiana benthamiana* seedlings at 3- to 5-leaf stage. When the inoculated plants are systemically infected, collect upper non-inoculated leaves to prepare target virus-containing homogenate.
2. Disinfect mortar and pestle with 95% ethanol, and place them on ice for 10 min.
3. Spread corundum powder on the newly expanded leaves of pepper plants infected by recombinant virus (*see* **Note 13**).
4. Collect symptomatic leaf infected by target virus, and slice it, and ground it in phosphate buffer in a ratio of 1:10 (g/mL).
5. Use the inoculum-dipped forefinger to slightly rub the newly expanded leaf of pepper plants (pre-infected by recombinant virus) two times (from leaf bottom to top).
6. Place challenging-inoculated plants in a dark room for 10 min, and then rinse off the remaining sap and corundum powder with clean water.
7. Place challenging-inoculated plants in a dark room for 1–2 h.
8. Maintain pepper plants in a plant growth cabinet (16-h light, 25 °C and 8-h darkness, 23 °C, 70% relative humidity) (*see* **Note 2**).
9. Evaluate the resistance of pepper plants (pre-infected by recombinant virus) to target virus by symptom observation, as well as other molecular methods such as RT-PCR and Western blot.

4 Notes

1. Soak pepper seeds for at least 24 h in sterilized double-distilled water. This step facilitates to break seed dormancy, increasing seed germination efficiency.
2. Based on our best experience, the efficiency of heterologous protein expression by using PVMV-based vector is enhanced in pepper plants in the following conditions: 16-h light, 25 °C and 8-h darkness, 23 °C, 70% relative humidity.
3. *Bsa* I is a commonly used type IIS restriction enzyme. It recognizes a 6-bp motif "GGTCTC". However, its cleavage site (4-bp) is outside the recognition motif, and 1 bp away from the recognition motif. All original motifs recognized by *Bsa* I in pPVMV-GoldG have been destroyed by synonymous mutations [16].
4. It is critical to check whether the *CP* cistron of target virus contains *Bsa* I recognition motif(s); otherwise, it will be unsuccessful to perform Golden Gate cloning to construct PVMV-based recombinant vector.
5. The 5′-termini of both primers must contain *Bsa* I recognition motif "GGTCTC" together with several protective bases such as "ATATA" (Fig. 1b). The sticky ends of digested PCR products ("CAGGT" and "TCGTC") could be ligated with *Bsa* I-treated PVMV vector.
6. The treatment "10 min at 80°C" will inactivate all enzymes in the reaction. Omitting this step would result in self-ligation to form empty vector—pPVMV-GoldG.
7. pPVMV-GoldG encoded a kanamycin-resistant gene. Therefore, kanamycin with proper concentration should be used for screening positive clones.
8. The pair of primers PV-P1-F/PV-HCPro-R locate at the upstream and downstream of inserted sequence in the resulting recombinant vector. Thus, the pair of primers should be used to screen positive clones.
9. The obtained plasmids should be kept in −20 °C for 6-month use, or −80 °C for long-time storage.
10. We recommend a further identification of the obtained recombinant plasmids by restriction digestion with *Mlu* I and *Bam*H I, before sending out for Sanger sequencing.
11. Do not warm competent cells above 4 °C. Otherwise, it will significantly decrease transformation efficiency.
12. The cell suspension with a higher value of OD_{600} will trigger inoculated leaf death.

13. It is necessary to check all plants pre-inoculated with recombinant clone showing green fluorescence signals in upper non-inoculated leaves, an indication of viral systemic infection. Usually, it will take 15 days for the systemic infection in pepper. At this time point, challenging inoculation with target virus would achieve better resistance efficiency.

Acknowledgments

This work was supported by grants from the National Natural Science Foundation of China (31860487), and the Hainan Provincial Natural Science Foundation of China (320RC717).

References

1. Kenyon L, Kumar S, Tsai WS, Hughes JDA (2014) Virus diseases of peppers (Capsicum spp.) and their control. Adv Virus Res 90: 297–354
2. Kumari N, Sharma V, Patel P, Sharma PN (2023) Pepper mild mottle virus: a formidable foe of capsicum production-a review. Front Virol 3:1208853
3. Ali A (2024) Overview of RNA viruses infecting Capsicum species. In: Pepper Virome. Academic, pp 67–98
4. Rosa C, Kuo YW, Wuriyanghan H, Falk BW (2018) RNA interference mechanisms and applications in plant pathology. Annu Rev Phytopathol 56:581–610
5. Voloudakis AE, Kaldis A, Patil BL (2022) RNA-based vaccination of plants for control of viruses. Annu Rev Virol 9:521–548
6. Kothari SL, Joshi A, Kachhwaha S, Ochoa-Alejo N (2010) Chilli peppers-a review on tissue culture and transgenesis. Biotechnol Adv 28:35–48
7. Abrahamian P, Hammond RW, Hammond J (2020) Plant virus-derived vectors: applications in agricultural and medical biotechnology. Annu Rev Virol 7:513–535
8. Cody WB, Scholthof HB (2019) Plant virus vectors 3.0: transitioning into synthetic genomics. Annu Rev Phytopathol 57:211–230
9. Li F, Wang A (2019) RNA-targeted antiviral immunity: more than just RNA silencing. Trends Microbiol 27:792–805
10. Brunt AA, Kenten RH (1971) Pepper veinal mottle virus-a new member of the potato virus Y group from peppers (Capsicum annuum L. and C. frutescens L.) in Ghana. Ann Appl Biol 69:235–243
11. Chung BYW, Miller WA, Atkins JF, Firth AE (2008) An overlapping essential gene in the Potyviridae. Proc Natl Acad Sci USA 105: 5897–5902
12. Mäkinen K, Aspelin WWE, Pollari M, Wang L (2023) How do they do it? The infection biology of potyviruses. Adv Virus Res 117:1–79
13. Majer E, Salvador Z, Zwart MP, Willemsen A, Elena SF, Daròs JA (2014) Relocation of the NIb gene in the tobacco etch potyvirus genome. J Virol 88:4586–4590
14. Pasin F (2021) Oligonucleotide abundance biases aid design of a type IIS synthetic genomics framework with plant virome capacity. J Biotechnol 16:e2000354
15. Hu W, Qin L, Yan H, Miao W, Cui H, Liu W (2020) Use of an infectious cDNA clone of pepper veinal mottle virus to confirm the etiology of a disease in Capsicum chinense. Phytopathology 110:80–84
16. Deng C, Hu W, Shen W, Li F, Dai Z, Cui H (2025) A point mutation in the pepper Veinal mottle virus 6K1 protein yields a stable attenuated strain for engineering virus resistance in pepper plants. Plant Dis 109:1459–1469
17. Hu W, Deng C, Qin L, Liu P, Wang L, Wang X et al (2025) A conserved lysine/arginine-rich motif is essential for the autophagic degradation of potyviral 6K1 protein and virus infection. J Virol 99:e0218324

Chapter 5

Construction of Attenuated Viruses by Changing the UpA Dinucleotide Frequency

Shuaifa Chang, Yong Li, Yixin Huang, Bo Song, Adrian Valli, Xiaoyun Wu, Xiaofei Cheng, and Weiqin Ji

Abstract

Cross protection with an attenuated virus is a promising strategy for controlling viral diseases in crops that lack resistance resources. Attenuated viruses are usually obtained from naturally avirulent strains or through mutagenesis. Studies have shown that increasing the UpA dinucleotide content in the genome without altering the amino acid encoding sequence causes the loss of virulence. This chapter describes a detailed protocol for developing attenuated virus by increasing the UpA dinucleotide content, using turnip mosaic virus as a case study. This approach enables the rapid acquisition of stably, inheritable attenuated virus strains.

Key words UpA, Cross-protection, Attenuated strain, TuMV, UpA dinucleotide

1 Introduction

Plant viral diseases are a significant biotic stress that limits agricultural production, necessitating green and efficient control strategies [1]. The prospect of "cross-protection"based on attenuated strains is promising and represents a classic biological control strategy for plant viruses [2]. This involves pre-inoculating healthy plants with a weakly pathogenic viral strain that has been artificially selected or genetically modified to induce a systemic disease resistance state. This enables the plants to effectively resist subsequent infections by virulent strains of the same virus [3]. However, the existing artificial cross-protection technologies primarily have two main limitations, namely, the acquisition of natural attenuated virus strains and the biosafety and genetic stability of these virus strains [4, 5]. Studies have shown that the CpG and UpA dinucleotide ratios are under-represented in many RNA viruses that infect eukaryotic organisms, and that an engineered virus with an elevated CpG or UpA

Guanwei Wu and Xiaofei Cheng (eds.), *Plant Antiviral Resistance: Methods and Protocols*, Methods in Molecular Biology, vol. 3039, https://doi.org/10.1007/978-1-0716-5300-5_5,

dinucleotide ratio may trigger host antiviral responses and therefore attenuate viral pathogenicity [6–8].

In this chapter, we describe an efficient method for creating attenuated virus by enhancing the UpA content, using the turnip mosaic virus (TuMV) as an example. We successfully screened out strains with significantly reduced pathogenicity and a stable genetic foundation, providing a novel technology for the rational design of plant virus vaccines.

2 Materials

2.1 Plant Materials

Nicotiana benthamiana used in this experiment is preserved by the Laboratory of Molecular Plant Virology in the Northeast Agricultural University.

2.2 Viruses and Bacteria

1. *Escherichia coli* strain DH5α.
2. *Agrobacterium tumefaciens* strain GV3101.
3. TuMV isolate UK1 infectious clone expressing a free YFP between P1 and HcPro cistrons (pCB301-TuMV-YFP).

2.3 Reagents

1. Kanamycin (Kana) and Rifampicin (Rif).
2. ClonExpress II One Step Cloning Kit.
3. FastPure Plasmid Mini Extraction Kit.
4. FastPure Gel DNA Mini Extraction Kit.
5. RNA Extraction Kit.
6. Magnesium chloride ($MgCl_2$).
7. 2-Morpholinoethanesulphonic acid.

2.4 Laboratory Equipment

1. Incubator Shaker.
2. Plant Tissue Grinder.
3. Plant Growth Chamber.
4. Eporator® Electroporator.
5. pH Meter.
6. Real-time PCR System.
7. Dual-wavelength fluorescent protein laser emitter.

2.5 Buffers

1. Liquid LB medium: Dissolve 10 g of peptone, 5 g of yeast extract, and 10 g of NaCl in distilled water. Adjust the pH to 7.0 and bring the final volume to 1 L. Autoclave at 121 °C for 15 min.
2. Solid LB medium: Dissolve 10 g of peptone, 5 g of yeast extract and 10 g of NaCl in distilled water. Then, add 15 g of agar and bring the final volume to 1 L. Adjust the pH to 7.0. Autoclave at 121 °C for 15 min.

3. Acetosyringone (AS): Weigh out 0.1962 g of AS and dissolve it in DMSO to reach a final volume of 10 mL. Immediately aliquot the stock solution into sterile 1.5 mL microcentrifuge tubes using a precision pipette (e.g., 10 μL per tube). Store all the aliquots at −20 °C in complete darkness.
4. 1 M $MgCl_2$: Accurately weigh out 10.165 g of $MgCl_2 \cdot 6H_2O$ powder and dissolve it directly in 30 mL of ddH_2O. Stir until the solution is completely clear. Adjust the final volume to 50 mL with ddH_2O. Sterilize by filtration through a 0.22 μm membrane filter and store at 4 °C.
5. 1 M MES: Accurately weigh out 9.76 g of MES powder and dissolve it in approximately 40 mL of ddH_2O. Stir at room temperature until completely dissolved. Using a calibrated pH meter, adjust the pH to 5.6 by slowly adding NaOH solution drop by drop. Bring the final volume up to 50 mL with ddH_2O. Sterilize by filtering through a 0.22 μm membrane filter. Divide into sterile tubes and store at 4 °C, protected from light.
6. Infiltration buffer: Pipette 500 μL of 1 M $MgCl_2$ solution and 500 μL of 1 M MES (pH 5.6) solution into a suitable container. Add ddH_2O to make up to approximately 45 mL, mix well, and then top up to 50 mL with ddH_2O. Prepare fresh before use or store short-term at 4 °C.

3 Methods

3.1 Analysis of Viral Genome Structure

1. Download the genome sequence of TuMV isolate UK1 from the NCBI GenBank database (Accession number: AB194792).
2. Analyze the RNA structure of TuMV genome on the official page of RNA fold Web Server (http://rna.tbi.univie.ac.at/cgi-bin/RNAWebSuite/RNAfold.cgi) (*see* **Note 1**).
3. Paste the prepared TuMV genome sequence in FASTA format or as a pure sequence into the input box on the page (*see* **Note 2**).
4. Save the results of RNA secondary structure diagram and minimum free energy and in formats such as PNG, SVG, or EPS (*see* **Note 3**).
5. Analyze putative reverse open reading frames in the viral genome at the NCBI ORF Finder (https://www.ncbi.nlm.nih.gov/orffinder/).
6. Set the genetic code table as the standard genetic code and the ORF length threshold of a minimum of 30 nucleotides (10 amino acids) (*see* **Notes 4** and **5**).
7. Download the reverse ORF prediction result to your local machine.

8. Import the above results into SnapGene to select regions that have no special secondary structure and no reverse open reading frames. The region selected in this study is TuMV (6876–7876 bp).
9. Save the genomic region as a separate file.

3.2 Engineering Viral Genome with Increased UpA Content

1. Download the most recent version of Perl (e.g., Strawberry Perl at https://strawberryperl.com/) based on your computer system (*see* **Notes 6** and 7).
2. Double-click the downloaded file in msi format and install it using the default settings (*see* **Note 8**).
3. Press Win + R keys, type cmd, and press Enter (*see* **Note 9**).
4. Use the *perl -v* command in the command window to verify whether Perl has been successfully installed (*see* **Note 10**).
5. Convert the target viral genome sequence to a fasta format file, and save it to your local machine.
6. Copy our Perl script and save to your computer as UpA.pl (*see* **Note 11**).
7. Execute the Perl script using the *cd your path*\UpA.pl command in the terminal (*see* **Note 12**).
8. Save the output to your local machine.
9. Send the sequence with increased UpA content for synthesis.

3.3 Vector Construction

1. Design primers to amplify the synthesized TuMV fragment with increased UpA content. The required primers are as follows: TuMV-UP-F (agactagtacaataatgcca GTAGAGAATAGT CAGTTTTGGAAAC) TuMV-UP-R (caaccacaataaaactttgt CATACCTACGCTCCAAGGAGCTCTT) (*see* **Note 13**).
2. Design primers to linear the pCB301-TuMV-YFP vector excluding the 6876–7876 bp region. The required primers are as follows: pCB301-TuMV-ZT-F(acaaagttttattgtggttgg-gatc) pCB301-TuMV-ZT-R(tggcattattgtactagtctcagagactat).
3. Perform PCR to amplify the synthesized TuMV fragment and pCB301-TuMV-YFP (*see* **Notes 14** and **15**).
4. Use ClonExpress II One Step Cloning Kit to insert the synthesized TuMV fragment into pCB301-TuMV-YFP to generate the pCB301-TuMV-GFP/UpA_high construct (*see* **Note 16**).
5. Transform the ligation reaction mix into *E. coli* (*see* **Note 17**).
6. Select correct clones by colony PCR with primers for the synthesized TuMV fragment.
7. Extract the plasmid (*see* **Note 18**).
8. Confirm the insertion by Sanger sequencing (*see* **Note 19**).

9. Transform pCB301-TuMV-GFP/UpA_high into *A. tumefaciens* strain GV3101 (*see* **Note 20**).
10. Culture the correct *A. tumefaciens* colonies in liquid LB medium containing Kana and Rif antibiotics overnight at 28 °C (*see* **Note 21**).

3.4 Agrobacterium Infiltration

1. Centrifuge at 6000 rpm for 1 min at room temperate to collect the bacteria.
2. Discard the supernatant and resuspend the bacterial pellets in 1 mL of infiltration buffer.
3. Repeat **steps 1** and **2** one more time.
4. Add AS to a final concentration of 10 mM.
5. Adjust the bacterial optical density at 600 nm (OD_{600}) of 0.4.
6. Let the bacterial solution at room temperature for at least 1 h.
7. Infiltrate the bacteria solution into the leaves of 4-week-old *N. benthamiana* seedlings with a 1 mL needleless syringe (*see* **Note 22**).
8. Place the plants back to a growth chamber for virus symptom development.
9. Record the virus symptoms each day and take the photos at desired time.
10. Collect the systemic symptomatic leaves for analyzing.

3.5 Expected Results

N. benthamiana plants infiltrated with *A. tumefaciens* harboring the pCB301-TuMV-GFP vector display typical viral symptoms, such as yellowing and mottling of the leaves and a dwarf and stunted appearance. In contrast, plants infiltrated with *A. tumefaciens* harboring the pCB301-TuMV-GFP/UpA_high vector resemble the mock-treat plants (Fig. 1a). RT-qPCR analysis show that *N. benthamiana* plants infiltrated with *A. tumefaciens* harboring the pCB301-TuMV-GFP vector accumulates high levels of TuMV genomic RNA. In contrast, only trace amounts of viral genomic RNA are detected in the plants infiltrated with pCB301-TuMV-GFP/UpA_high (Fig. 1b). Taken together, these data indicate that the pathogenicity of the engineered TuMV is greatly reduced.

4 Notes

1. If the sequence length exceeds 7500 bp, it can be analyzed in two parts.
2. Under normal circumstances, you can start the analysis by keeping the default parameters on the page. These default parameters are suitable for most cases.

Fig. 1 Phenotypes of *N. benthamiana* seedlings infected by wild-type TuMV-GFP and TuMV-GFP/UpA_high. (**a**) Phenotype of *N. benthamiana* plants under ultra violet at 7 dpi. (**b**) RT-qPCR assay of the accumulation of TuMV genome in the systemic leaves. *** indicates $p < 0.001$ in the pairwised Student T-test

3. The lower minimum free energy value, the more stable the predicted secondary structure is likely to be.
4. The standard genetic code is the most appropriate one for most of the cases. However, a specific viral code table may be needed if it is a viral sequence.
5. The default is 75 minimum nucleotides (25 amino acids). Here, we choose a minimum of 30 nucleotides (10 amino acids) to avoid missing shorter potential ORFs.
6. Strawberry Perl is more suitable for beginners.
7. Please select suitable version of Strawberry Perl based on your computer system. For example, select the Strawberry Perl for Windows (64-bit version) for most Windows users.
8. For Mac and Linux users, please use command to install Strawberry Perl as instructed.
9. For Mac and Linux users please open "Applications" → "Utilities" → "Terminal" and enter: perl -v and press Enter.
10. If you see version information (such as "This is perl 5, version…"), it means Perl is installed successfully. If it displays

"Not recognized as an internal or external command", it means you need to reinstall it.

11. Script list is as follows:

```
#!/usr/bin/env perl
=pod
###### 2024-08-19 20:41 - Yong Li
This script is to increase UpA content of CDSs by substi-
tution of synonymous codons.
Examples:
 Priority 1: Asp: GAC -> GAT
 Priority 2: ArgIle: CGAATC -> CGTATC
 Priority 3: ThrSer: ACATCC ->ACTAGT
=cut
use Time::Interval;
my $thisScript = $0;
$thisScript = $1 if ( $0 =~ /\cr([^\cr]+)$/ || $0 =~ /\/([^
\/]+)$/);
if ( @ARGV != 2 || $ARGV[0] eq "-h" || $ARGV[0] eq "-help"
|| $ARGV[0] eq "--help") {
 print "\n$0:\n";
  print "Usage:./$thisScript in_CDS_fasta_file out_fasta_-
file.\n";
 print "This script is to increase UpA content of CDSs by
substitution of synonymous codons.\n\n";
 exit(1);
}
print "\nRunning $thisScript". ( defined( $ARGV[0]) ? " for
$ARGV[0] ": " "). "now...\n";
my $start_time = localtime(time);
my (%seq_c, $id, $seq);
$seq = '';
open( IN, "<$ARGV[0]") or die "$!\n";
while ( defined( my $in = <IN>)) {
 chomp($in);
 if ($in =~ /^>/){
  if ($seq ne ''){
    $seq_c{$id} = $seq;
    #print "ID: $id\nSeq: $seq\n";
 }
  $in =~ s/\s.*//;
  $id = $in;
  $seq = '';
 }
 else {
  $in =~ s/\s+//g;
  $in =~ tr/acgt/ACGT/;
  $in =~ s/\d+//g;
  $seq.= $in;
```

```
 }
}
if ($seq ne ''){
 $seq_c{$id} = $seq;
 #print "ID: $id\nSeq: $seq\n";
}
close(IN);
open( OUT, ">$ARGV[1]") or die "$!\n";
foreach my $key (keys %seq_c) {
 my $value = $seq_c{$key};
 my $next_a = 0;
 for (my $i=0; $i<length($value); $i+=3){
   if ($i+3 <= length($value) && (substr($value, $i+3, 1)
eq "A" || substr($value, $i+3, 3) =~ /CG[ACGT]/ || substr
($value, $i+3, 3) =~ /TC[ACGT]/)){
    $next_a = 1;
 }
 else {
    $next_a = 0;
 }
 my $codon = substr($value, $i, 3);
 if ($codon eq "TTG"){
    substr($value, $i, 3, "TTA");
 }
 elsif ($codon eq "GTT" || $codon eq "GTC" || $codon eq
"GTG"){
    substr($value, $i, 3, "GTA");
 }
 elsif ($codon eq "CTT" || $codon eq "CTC" || $codon eq
"CTG"){
    substr($value, $i, 3, "CTA");
 }
 elsif ($codon eq "ATT" || $codon eq "ATC"){
    substr($value, $i, 3, "ATA");
 }
 elsif ($codon eq "TGA"){
    substr($value, $i, 3, "TAG");
 }
 elsif ($next_a){
    if ($codon eq "GCA" || $codon eq "GCC" || $codon eq
"GCG"){
     substr($value, $i, 3, "GCT");
 }
 elsif ($codon eq "TCA" || $codon eq "TCC" || $codon eq
"TCG"){
    substr($value, $i, 3, "TCT");
 }
 elsif ($codon eq "GGA" || $codon eq "GGC" || $codon eq
"GGG"){
```

```
        substr($value, $i, 3, "GGT");
    }
    elsif ($codon eq "CGA" || $codon eq "CGC" || $codon eq
  "CGG"){
        substr($value, $i, 3, "CGT");
    }
    elsif ($codon eq "CCA" || $codon eq "CCC" || $codon eq
  "CCG"){
        substr($value, $i, 3, "CCT");
    }
    elsif ($codon eq "ACA" || $codon eq "ACC" || $codon eq
  "ACG"){
        substr($value, $i, 3, "ACT");
    }
    elsif ($codon eq "TTC"){
        substr($value, $i, 3, "TTT");
    }
    elsif ($codon eq "TAC"){
        substr($value, $i, 3, "TAT");
    }
    elsif ($codon eq "GAC"){
        substr($value, $i, 3, "GAT");
    }
    elsif ($codon eq "CAC"){
        substr($value, $i, 3, "CAT");
    }
    elsif ($codon eq "AGC"){
        substr($value, $i, 3, "AGT");
    }
    elsif ($codon eq "AAC"){
        substr($value, $i, 3, "AAT");
     }
    }
    elsif ($codon =~ /CG[ACGT]/){
       substr($value, $i, 3, "AGG");
    }
    elsif ($codon =~ /TC[ACGT]/){
       substr($value, $i, 3, "AGT");
    }
  }
    print OUT "$key"."_UpA\n"."$value\n";
}
close(OUT);
my $end_time = localtime(time);
my $elapsed_time = getInterval( $start_time, $end_time,
  'String' => True);
print "... $thisScript running finished, used time: $elap-
  sed_time.\n\n";
```

12. Please replace *your path* with the exact path of the UpA.pl saved. Such as: C:\desktop\UpA.pl.
13. The primers contain a 20-nucleotide overlapping with neighboring sequence.
14. The PCR products should be analyzed by electrophoresis on a 1% agarose gel and then purified using a DNA gel extraction kit.
15. At the end of amplification for pCB301-TuMV-YFP, 1 μL of restriction enzyme *Dpn* I should be added and incubated at 37 °C for an additional of 1 h to cleave the template plasmid.
16. Please follow the manufacturer instructions for the reaction.
17. Either heat shock or electroporation can be used for transformation.
18. Do not forget to add RNase A to Buffer P1 for the plasmid extraction to avoid RNA contamination.
19. Make sure the entire synthesized TuMV fragment including the overlapping region are completely sequenced and the codons in frame and putative amino acid sequence is identical to the original TuMV.
20. Colony PCR should be used to select correct *A. Tumefaciens* colonies.
21. The bacterial culture can be stored at −80 °C after mixing with an equal volume of 50% glycerol for long-term storage. Glycerol stock should not be repeatedly frozen and thawed.
22. *A. tumefaciens* harboring pCB301-TuMV-GFP should be included for comparison.

References

1. Tatineni S, Hein GL (2023) Plant viruses of agricultural importance: current and future perspectives of virus disease management strategies. Phytopathology 113:117–141
2. Tomitaka Y, Shimomoto Y, Ryang BS et al (2024) Development and application of attenuated plant viruses as biological control agents in Japan. Viruses 16:517
3. Loebenstein G, Carr JP (eds) (2006) Natural resistance mechanisms of plants to viruses. Springer, Berlin
4. Sato H, Seo N, Nakamura S (2012) Attenuated mutants of lily mottle virus obtained by tissue culture of lily bulb. Hortic Res 11:113–120
5. Hernando Y, Aranda MA (2023) Cross-protection against Pepino mosaic virus, more than a decade of efficient disease control. Ann Appl Biol 184:174–182
6. González de Prádena A, Sánchez Jimenez A, San León D et al (2020) Plant virus genome is shaped by specific dinucleotide restrictions that influence viral infection. MBio 11:e02818–e02819
7. Cheng X, Virk N, Chen W et al (2013) CpG usage in RNA viruses: data and hypotheses. PLoS One 8:e74109
8. Odon V, Fros JJ, Goonawardane N et al (2019) The role of ZAP and OAS3/RNAseL pathways in the attenuation of an RNA virus with elevated frequencies of CpG and UpA dinucleotides. Nucleic Acids Res 47:8061–8083

Chapter 6

Mild Strain Cross-Protection in Passion Fruit

Linxi Wang, Hongguang Cui, and Zhaoji Dai

Abstract

Cross-protection is a phenomenon in which a plant prior infected with a mild viral strain can be protected from subsequent infection by severe strains of the same or closely related viruses. Since its discovery in the 1920s, this phenomenon has become an important strategy for controlling viral diseases in plants. Telosma mosaic virus (TelMV, a species of *Potyvirus*) is an emerging pathogen that threatens passion fruit production in subtropical and tropical regions. Here, we present a protocol for screening mild TelMV strains and their application in protecting passion fruit from severe TelMV infections. This provides an efficient TelMV management strategy and a framework for cross-protection implementation against other plant viruses.

Key words Cross protection, TelMV, Mild strain, Potyvirus, Passion fruit

1 Introduction

Cross-protection is a phenomenon in which plants that are pre-infected with a mild strain of a virus can resist subsequent infections by a severe strain of the same virus or closely related viruses. [1, 2]. This phenomenon was first observed by McKinney in 1929 during his research on tobacco mosaic virus (TMV). He discovered that tobacco plants infected with a TMV strain that caused green mosaic symptoms became immune to another TMV strain that induced yellow mosaic symptoms. Since then, cross-protection has emerged as an effective and environmentally friendly method for preventing various plant virus diseases. It has been successfully applied in controlling viruses such as cucumber green mottle mosaic virus (CGMMV), pepino mosaic virus (PepMV), citrus tristeza virus (CTV), papaya ringspot virus (PRSV), and zucchini yellow mosaic virus (ZYMV) [3–6]. A key step in implementing cross-protection is the screening of mild or attenuated strains of the virus. Currently, the main methods for obtaining mild or attenuated strains include natural selection, chemical or physical mutagenesis, high or low-temperature treatment, the use

Guanwei Wu and Xiaofei Cheng (eds.), *Plant Antiviral Resistance: Methods and Protocols*, Methods in Molecular Biology, vol. 3039, https://doi.org/10.1007/978-1-0716-5300-5_6,

of satellite RNA, and site-directed mutagenesis through reverse genetics technology [7–11].

Passion fruit (*Passiflora edulis*) is a perennial evergreen vine plant that grows mainly in tropical and subtropical regions. It is renowned for its unique flavor, as well as its rich vitamins and nutrients. However, viral diseases became the major factor restricting the healthy development of passion fruit production. So far, more than 40 species of viruses that infect passion fruit have been identified. Among them, telosma mosaic virus (TelMV) is regarded as one of the primary risks to passion fruit production [12, 13]. Infected plants exhibit symptoms such as leaf mosaic, distortion, and uneven fruit coloring [14].

TelMV belongs to the genus *Potyvirus*, which is the largest group of plant RNA viruses. TelMV encodes eleven proteins, with the helper component proteinase (HC-Pro) being the second gene product at the amino terminus of the genome. HC-Pro is a multifunctional protein that plays roles in aphid transmission, viral replication, viral movement, protease activity, and gene silencing suppression. This protein harbors a highly conserved motif, FRNK, in its central region [15, 16]. Research has shown that mild or attenuated strains generated by mutations in Arginine (R) residues within the FRNK motif can provide cross-protection against infection by severe strains, including turnip mosaic virus (TuMV), ZYMV, and sugarcane mosaic virus (SCMV) [8, 17–19].

This protocol outlines the process for generating mild strains of TelMV. A summary of the workflow is provided in Fig. 1. In this study, we created two-point mutants by modifying the Arginine residue within the FRNK motif of the HC-Pro protein to either aspartic acid or glutamic acid. Our results showed that these point mutants resulted in mild strains that could effectively provide cross-protection against subsequent challenges from severe TelMV in passion fruit plants.

2 Materials

2.1 Plant Material and Virus Infectious Clones

1. Tobacco (*Nicotiana benthamina*) and passion fruit (*Passiflora edulis*) seedlings are grown in pots and placed in a plant growth chamber with light for 16 h (24 °C) and darkness for 8 h (22 °C) at a relative humidity of 65%.
2. Virus infectious clones: (1) TelMV infectious clone, pPasFru (henceforth WT). This virus infectious clone is used for the generation of TelMV HC-Pro mutants, and (2) GFP-tagged TelMV infectious clone, pPasFru-GFP (henceforth TelMV-GFP). It is used for the challenge inoculation [20].

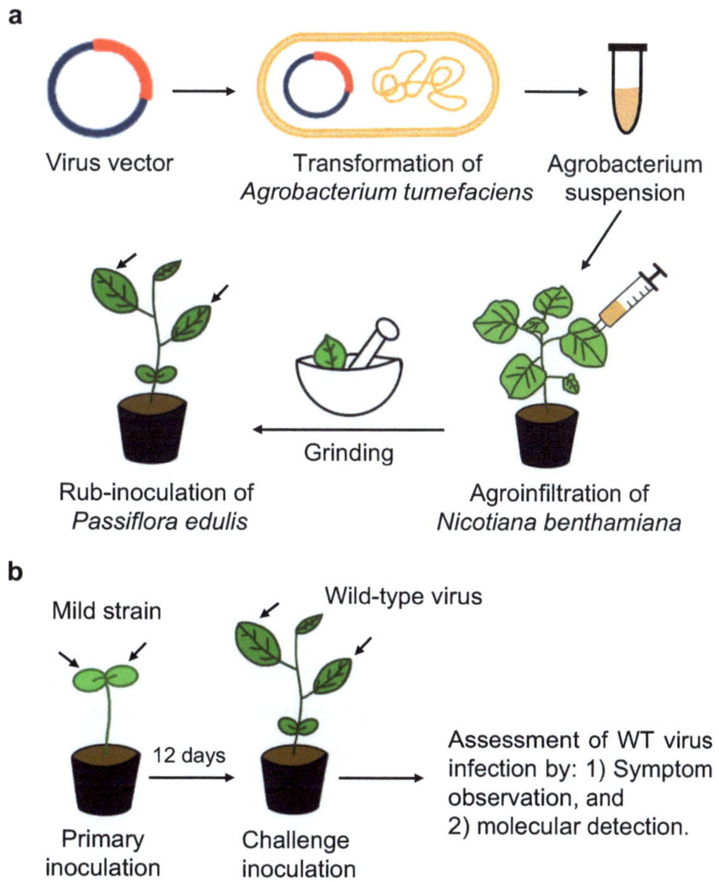

Fig. 1 Flowchart of the experimental steps for cross-protection in passion fruit plants. (**a**) A schematic diagram of inoculating the virus clone into passion fruit plants. Arrows indicate the leaves for inoculation. (**b**) A schematic diagram of challenge inoculation assay in passion fruit. Arrows indicate the leaves for rub-inoculation

2.2 Generation of TelMV HC-Pro Mutants

1. Sterilized plastic material (tubes, tips, etc.).
2. TelMV infectious clone pPasFru [20].
3. Phusion High-Fidelity DNA Polymerase.
4. Taq DNA polymerase mix.
5. Gel Extraction Kit.
6. *Pst*I and *Sal*I restriction enzyme.
7. Cycle-Pure Kit.
8. T4 DNA ligase Kit.
9. *Escherichia coli* strain DH10B chemically competent cells.
10. Luria Bertani medium (LB): 5 g L^{-1} Yeast extract, 10 g L^{-1} Tryptone, 10 g L^{-1} NaCl, pH 7.0. Autoclave at 121 °C for 20 min. For making solid media: add Agar (15 g L^{-1}) and autoclave at 121 °C for 20 min.

11. 100 mg/mL Kanamycin stock solution in water. Sterilize by filtration through a 0.22 μm filter. Store at −20 °C.
12. TIANprep Rapid Mini Plasmid Kit.

2.3 Agroinfiltration and Rub-Inoculation

1. *Agrobacterium tumefaciens* strain GV3101 electrocompetent cells.
2. 20 mg/mL Rifampicin stock solution in DMSO. Sterilize by filtration through a 0.22 μm filter. Store at −20 °C.
3. Agroinfiltration buffer: 10 mM $MgCl_2$, 10 mM MES, 200 mM acetosyringone, pH 5.7.
4. Sterile syringe.
5. Rub-inoculation phosphate buffer: 1 mL Rub-inoculation phosphate buffer consists of 490 μL of 10 mM Na_2HPO_4 and 510 μL of 10 mM KH_2PO_4.
6. Carborundum powder (400 mesh).
7. RNAprep Pure Polysaccharide Polyphenol Plant Total RNA Extraction Kit.

2.4 Equipment

1. Thermocycler.
2. Gel Electrophoresis Equipment.
3. Water bath.
4. Centrifuge.
5. Magnetic stirrer.
6. Vortexer.
7. 37 °C and 28 °C incubators and shakers.
8. Petri dishes (100×15 mm^2).
9. Electroporator.
10. Cuvette.
11. Mortar and Pestle.
12. Nanodrop.
13. pH meter.
14. Laminar flow hood.
15. Spectrophotometer.
16. Ultraviolet Lamp.

3 Methods

Mutate the Arginine at position 181 of HC-Pro into two negatively charged amino acids: aspartic acid (Asp, D), glutamic acid (Glu, E) by overlapping PCR.

3.1 Mutagenesis for Generating TelMV Mutants

3.1.1 Primer Design

1. Design flanking primers F1 and R1: completely match with the template DNA and include restriction enzyme sites for inserting the product into the vector. F1: AACTCTGAG**CTGCA**GAAACGA; R1: TGAGT**GTCGAC**CTCTTTCACCC. The restriction enzyme sites (*Pst*I and *Sal*I) are marked in bold. F: forward primer; R: reverse primer.
2. Design the mutation primers F2 and R2: carrying a point mutation, with a 15–25 bp overlapping sequence at the 5′ end of F2 and R2 (the mutation site is located in the center of the overlapping region). For example: mutating R (codon: AGG) into D (codon: GAT): F2: CTCAAGACTTTCGATAATAAGCGTTCCTCAAAAGCC; R2: GGCTTTTGAGGAACGCTTATTATCGAAAGTCTTGAG. The mutation sites are marked in gray.

3.1.2 Obtain the Fragment with the Mutation Site

1. Use the pPasFru plasmid as the template. Perform primary PCR with Phusion High-Fidelity DNA Polymerase. More specifically, amplify fragment I using primers F1/R2 and fragment II using primers F2/R1. Prepare each 50 μL reaction by mixing 41 μL sterilized distilled water, 5 μL 10× Phusion buffer, 1 μL 10 mM primer F, 1 μL 10 mM primer R, 1 μL 10 mM dNTPs, 0.5 μL template DNA, and 0.5 μL Phusion High-Fidelity DNA Polymerase. Run thermal cycling at 98 °C for 30 s, 30 cycles of 98 °C for 10 s, 52 °C for 30 s, 68 °C for 1 min, final extension at 68 °C for 10 min (*see* **Note 1**).
2. Analyze the PCR product by 1% agarose gel electrophoresis. A single and discrete band should be visualized, and make sure the size is correct.
3. Use the Gel Extraction Kit to purify the target band (fragments I and II).
4. Use the fragments I and II as the templates, and F1/ R1 as primers for obtaining fragment III by PCR. Purify the PCR product as described above.

3.1.3 Insert the Fragment into TelMV Infectious Clone (pPasFru) to Obtain the Recombinant Viral Mutants

1. Digest fragment III and TelMV infectious clone, pPasFru, with the restriction enzymes *Pst*I and *Sal*I according to the supplier's instructions.
2. Use the Cycle-Pure Kit to purify fragment III upon digestion.
3. Analyze the digested vector by electrophoresis on a 1% agarose gel to verify the proper linearization of the vector and purify it using the Gel Extraction Kit.
4. Perform ligation of the *Pst*I/*Sal*I-digested pPasFru vector with the *Pst*I/*Sal*I-digested fragment III. The ligation reaction is prepared by mixing 20 ng of plasmid DNA, with 20 ng of fragment, 1 μL of 10× ligation buffer, and 1 μL of T4 DNA ligase in a final volume of 10 μL. Incubate at 22 °C for 3 h and then put on ice (*see* **Note 2**).

5. Transform *E. coli* DH10B chemically competent cells with the ligation product. Firstly, mix 5 μL ligation product with 50 μL cells into a tube, and incubate on ice for 30 min. Secondly, heat shock is performed at 42 °C for 45 s, and immediately transfer the tube onto ice for 2 min. Thirdly, add 700 μL antibiotic-free LB liquid medium, and incubate with horizontal shaking (180 rpm) at 37 °C for 1 h. Lastly, plate the suspension on an LB agar medium plate containing 100 μg/mL of kanamycin, and incubate overnight at 37 °C.
6. Identification of positive colonies by PCR with the specific primers. Prepare the amplification reaction mixing 3 μL of sterilized distilled water, 0.5 μL of 10 mM primer F, 0.5 μL of 10 mM primer R, 1 μL template (colony dilution), and 25 μL of Taq DNA polymerase mix at a 50 μL final volume. Visualize the PCR product by 1% agarose gel electrophoresis.
7. Propagation of the positive colony: Inoculate the *E. coli* suspensions of the positive colony in 5 mL of LB liquid medium containing 100 μg/mL kanamycin. Incubate at 37 °C overnight with agitation (180 rpm) in an inclined position.
8. Harvest the cells (2 mL) by centrifugating at 12000 rpm for 2 min and extract the plasmid DNA of the bacterial culture using TIANprep Rapid Mini Plasmid Kit.
9. Double digestion by *Pst*I/*Sal*I and Sanger sequencing to verify the recombinant viral mutants.

3.2 Virus Inoculation of Plants

3.2.1 Transformation of A. tumefaciens

1. Transform *A. tumefaciens* strain GV3101 electrocompetent cells with pPasFru plasmid and the recombinant viral mutant plasmids. Mix 1 μL plasmid with 50 μL cells into a tube, transfer the mixture into a cuvette carefully, and electroporate at 2.4 kV.
2. After electroporation, add 700 μL of antibiotic-free LB liquid medium into the cuvette. Mix by pipette and transfer the culture to a 1.5-mL microcentrifuge tube. Incubate with horizontal shaking (180 rpm) at 28 °C for 2 h. Spread on an LB agar medium plate containing 100 μg/mL kanamycin and 20 μg/mL of rifampicin. Incubate at 28 °C for 48 h.
3. Identification of positive *A. tumefaciens* colonies by PCR with the specific primers: The amplification reaction is the same as described before (Subheading 3.1.3, **step 6**).
4. Select a positive colony, inoculate *A. tumefaciens* culture in 3 mL of LB liquid medium containing 100 μg/mL kanamycin and 20 μg/mL of rifampicin. Incubate at 28 °C for 24–36 h with agitation (180 rpm) in an inclined position.

3.2.2 Agroinfiltration of Nicotiana benthamiana

1. Pellet the cells (1 mL) by centrifuging at 5000 rpm for 3 min. Discard the supernatant.
2. Resuspend the pellet with 1 mL of Agroinfiltration buffer by pipetting.
3. Centrifuge for 3 min at 5000 rpm. Discard the supernatant. Repeat **steps 2** and **3**.
4. Resuspend the pellet with 1 mL of Agroinfiltration buffer by pipetting.
5. Adjust the optical density value of the Agrobacterium mixture to 0.5.
6. Infiltrate the Agrobacterium cells harboring viral mutants into leaf tissues of 3–4-week-old *N. benthamiana* by using a 1-mL syringe (Fig. 1a). The Agrobacterium cells harboring wild-type pPasFru (WT) and Agroinfiltration buffer (Mock) serve as the positive control and negative control, respectively (*see* **Note 3**).

3.2.3 Rub-Inoculation of Passiflora edulis

1. Dust the upper surface of the first and the second true leaves of *P. edulis* (three-leaf stage) with carborundum.
2. Sample the infiltrated *N. benthamiana* leaves (0.1 g) and mix with 1 mL of rub-inoculation phosphate buffer in the mortar using a pestle.
3. Rub-inoculate the sap onto the two dusted leaves of *P. edulis* (*see* **Note 4**).

3.3 Obtaining Mild Mutants

1. Continuously monitor the symptoms of *P. edulis* caused by each viral mutant after rub-inoculation.
2. The WT virus induces severe symptoms, including plant stunting, leaf mosaic, and small foliage, at 25 days post rub-inoculation (dpr). In contrast, mutants $R_{181}D$, and $R_{181}E$ do not show such symptoms in passion fruit plants, which is similar to Mock plants (Fig. 2a).
3. Collect the upper uninoculated leaves and subject to total RNA isolation using RNAprep pure polysaccharide polyphenol plant total RNA extraction kit according to the manual instructions (*see* **Note 5**).
4. Synthesize cDNA from 500 ng of RNA using reverse transcriptase following the manual instructions (*see* **Note 6**).
5. Amplify the fragment of the viral *coat protein* (*CP*) gene by PCR to detect the accumulation of virus at the mRNA level. The amplification reaction is the same as described before (Subheading 3.1.3, **step 6**). RT-PCR analysis of the systemic leaves shows that viral RNA can be detected in the systemic leaves of passion fruit plants inoculated by the mutants $R_{181}D$, and $R_{181}E$ (Fig. 2b).

Fig. 2 Effect of mutating R181 in the FRNK motif of HC-Pro on telosma mosaic virus (TelMV) infection in passion fruit. (**a**) Passion fruit plants infected with buffer (Mock), wild-type (WT) virus or HC-Pro mutant virus at 25 days post rub-inoculation (dpr). Scale bars, 2 cm. (**b**) RT-PCR analysis of viral RNA accumulation in the upper non-inoculated leaves of passion fruit at 25 dpr

6. Based on the symptom observation and RT-PCR results, two mild strains ($R_{181}D$, and $R_{181}E$) are obtained for the subsequent cross-protection study.

3.4 Challenge Inoculation

1. Rub-inoculate the sap prepared from $R_{181}D$ or $R_{181}E$-agroinfiltrated *N. benthamiana* leaves onto the cotyledons of *P. edulis* (one-leaf-stage; Fig. 1b), with ten passion fruit plants for each mutant (*see* **Note 7**).
2. At 12 dpr, challenge-inoculate the first and the second true leaves of *P. edulis* with the sap prepared from WT virus tagged with GFP (TelMV-GFP)-infected *N. benthamiana* leaves (Fig. 1b; *see* **Notes 8** and **9**).
3. Evaluate the cross-protection effect through symptom observation and molecular detection. Visual observation and GFP signal detection using UV lamp revealed that two HC-Pro mutants ($R_{181}D$, and $R_{181}E$) conferred efficient cross-protection against TelMV-GFP infection in passion fruit at 30 days post challenge (dpc; Fig. 3a). The accumulations of GFP at the mRNA level in the upper non-inoculated leaves of passion fruit were analyzed by RT-PCR (Fig. 3b; *see* **Note 10**).

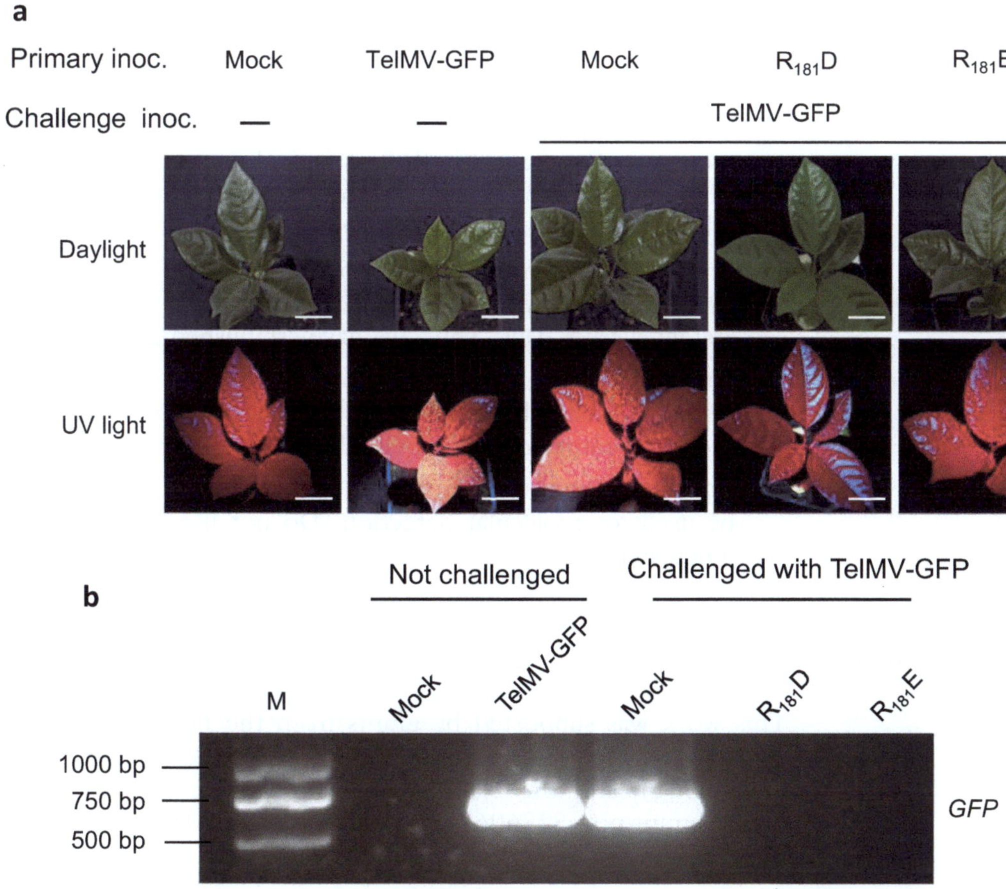

Fig. 3 Two telosma mosaic virus (TelMV) HC-Pro mutants could cross-protect against the severe strain (**a**) Two HC-Pro mutants confer efficient cross-protection against TelMV-GFP infection in passion fruit at 30 days post challenge (dpc), respectively. Scale bar, 3 cm. inoc., inoculation (**b**) RT-PCR analysis of *GFP* mRNA in the upper-noninoculated leaves of passion fruit at 30 dpc

4 Notes

1. To avoid introducing additional mutations, it is essential to use high-fidelity enzymes for PCR.
2. The ligation reaction can also be incubated at 16 °C overnight. The ligation product can be kept at −20 °C before transforming it into bacteria.
3. Make sure sufficient leaves are agroinfiltrated, as infiltrated leaves serve as the source of the subsequent rub-inoculation for passion fruit plants.
4. To prevent necrosis after rubbing, avoid using too much carborundum and ensure that you rub gently.

5. The leaves can also be stored at −80 °C and are subject to total RNA isolation at your convenience.
6. Confirm RNA quality before proceeding to cDNA synthesis.
7. Prior to challenge inoculation, it is essential to detect the presence of mutants in the systemic leaves using RT-PCR analysis. This step ensures that the *P. edulis* plants are systemically infected with the mutant virus.
8. The protection interval should be determined by the specific virus. Based on our observations, TelMV infects a leaf approximately every 5 days. Thus, we have established a 12-day protection interval.
9. Using a virus with a fluorescent tag for challenge inoculation allows for more convenient visual tracking of the infection.
10. The newly emerged leaf after the challenge-inoculation should be used for molecular detection. Do not use the challenge-inoculated leaf.

Acknowledgments

This work was supported by grants from the Hainan Provincial Natural Science Foundation (grant nos. 321QN181 and 322RC564), and the National Natural Science Foundation of China (grant nos. 32360651, 32102157, and 32372484).

References

1. Gonsalves D, Garnsey SM (1989) Cross-protection techniques for control of plant virus diseases in the tropics. Plant Dis 73: 592–597
2. Ziebell H, Carr JP (2010) Cross-protection: a century of mystery. Adv Virus Res 76:211–264
3. McKinney HH (1929) Mosaic diseases in the Canary Islands, West Africa and Gibraltar. J Agric Res 39:577–578
4. Folimonova SY (2013) Developing an understanding of cross-protection by citrus tristeza virus. Front Microbiol 4:76
5. Chewachong GM, Miller SA, Blakeslee JJ et al (2015) Generation of an attenuated, cross-protective Pepino mosaic virus variant through alignment-guided mutagenesis of the viral capsid protein. Phytopathology 105:126–134
6. Cheng HW, Lin TT, Huang CH et al (2023) Modification of papaya ringspot virus HC-pro to generate effective attenuated mutants for overcoming the problem of strain-specific cross protection. Plant Dis 107:1757–1768
7. Kosaka Y, Fukunishi T (1993) Attenuated isolates of soybean mosaic virus derived at a low temperature. Plant Dis 77:882–886
8. Lin SS, Wu HW, Jan FJ et al (2007) Modifications of the helper component-protease of zucchini yellow mosaic virus for generation of attenuated mutants for cross protection against severe infection. Phytopathology 97:287–296
9. Raja JAJ, Huang CH, Chen CC et al (2022) Modification of N-terminal FWKG-αH1 element of potyviral HC-Pro affects its multiple functions and generates effective attenuated mutants for cross protection. Mol Plant Pathol 23:947–965
10. Tuo D, Zhou P, Zhao G et al (2020) A double mutation in the conserved motifs of the helper component protease of papaya leaf distortion mosaic virus for the generation of a cross-protective attenuated strain. Phytopathology 110:187–193
11. Wang L, Shi W, Aziz A et al (2024) Mutating the arginine residue within the FRNK motif of telosma mosaic virus (TelMV) HC-Pro protein

attenuates viral infection and confers effective protection against TelMV in passion fruit (*Passiflora edulis*). Pest Manag Sci 80:5256–5265

12. Wu W, Ma F, Zhang X et al (2024) Research progress on viruses of *Passiflora edulis*. Biology 13:839
13. Zhang C, Jiang J, Chen S et al (2024) Telosma mosaic virus: an emerging plant RNA virus causing production loss in passion fruit across Asia. Plant Pathol 73:242–249
14. Yang K, Yan H, Song L et al (2018) Analysis of the complete genome sequence of a potyvirus from passion fruit suggests its taxonomic classification as a member of a new species. Arch Virol 163:2583–2586
15. Valli AA, Gallo A, Rodamilans B et al (2018) The HCPro from the Potyviridae family: an enviable multitasking helper component that every virus would like to have. Mol Plant Pathol 19:744–763
16. Shiboleth YM, Haronsky E, Leibman D et al (2007) The conserved FRNK box in HC-Pro, a plant viral suppressor of gene silencing, is required for small RNA binding and mediates symptom development. J Virol 81:13135–13148
17. Kung YJ, Lin PC, Yeh SD et al (2014) Genetic analyses of the FRNK motif function of Turnip mosaic virus uncover multiple and potentially interactive pathways of cross-protection. Mol Plant-Microbe Interact 27:944–955
18. Gal-On A (2000) A point mutation in the FRNK motif of the potyvirus helper component-protease gene alters symptom expression in cucurbits and elicits protection against the severe homologous virus. Phytopathology 90:467–473
19. Xu XJ, Li HG, Cheng DJ et al (2020) A spontaneous complementary mutation restores the RNA silencing suppression activity of HC-pro and the virulence of sugarcane mosaic virus. Front Plant Sci 11:549767
20. Gou B, Dai Z, Qin L et al (2023) A zinc finger motifin the P1 N terminus, highly conserved in a subset of potyviruses, is associated with the host range and fitness of Telosma mosaic virus. J Virol 97:e0144422

Chapter 7

A TurboID-Based Protocol for Efficient Interacting Proteins Identification Using Proximity Tagging Technology

Xinxin Fang, Jianping Chen, Fei Yan, and Guanwei Wu

Abstract

Understanding the dynamic protein–protein interactions (PPIs) is fundamental to deciphering cellular signaling pathways and biological processes. Traditional methods for identifying PPIs, such as yeast two-hybrid and co-immunoprecipitation, often lack spatiotemporal resolution and may miss weak or transient interactions. Proximity-dependent biotinylation (PDB) coupled with mass spectrometry (MS) has emerged as a powerful technique to map the spatial proteome in living cells. Among various engineered biotin ligases, Turbidite-Identification (TurboID) stands out for its unprecedented catalytic efficiency, enabling rapid biotinylation of neighboring proteins within minutes. This chapter provides a detailed, step-by-step protocol for conducting a TurboID-based proximity labeling experiment to identify potential interacting proteins of a target protein of interest.

Key words TurboID, *Nicotiana benthamiana*, Protein–protein interaction

1 Introduction

The precise elucidation of protein-protein interactions is central to uncovering the molecular mechanisms underlying biological processes. Traditional interaction mapping methods, such as yeast two-hybrid and co-immunoprecipitation, present several challenges when applied to plant systems, including limitations like mislocalization of proteins, poor stability of interactions, and difficulty in capturing transient interactions [1, 2].

Proximity-dependent biotinylation (PDB) technologies have revolutionized the mapping of protein interactions within the living, native environment of a cell. This technique relies on the genetic fusion of an engineered biotin ligase (e.g., TurboID) to a target "bait" protein [3–5]. Upon the addition of biotin, the enzyme is activated and generates a highly reactive biotin-AMP intermediate, which covalently labels lysine residues on neighboring proteins within a radius of approximately 10 nm [6]. These

Guanwei Wu and Xiaofei Cheng (eds.), *Plant Antiviral Resistance: Methods and Protocols*, Methods in Molecular Biology, vol. 3039, https://doi.org/10.1007/978-1-0716-5300-5_7,

biotinylated proteins can then be efficiently isolated using streptavidin beads and identified via liquid chromatography-tandem mass spectrometry (LC-MS/MS), thereby revealing the potential interaction network of the bait protein.

Among various proximity-labeling enzymes, TurboID stands out due to its exceptional catalytic efficiency, enabling robust labeling in as little as 10 min. This represents a significant advantage over first-generation enzymes like BioID, which required 18–24 h [7–10]. This characteristic makes TurboID particularly suitable for studying rapidly responding biological processes, such as plant immune signaling, and effectively reduces background noise associated with prolonged labeling.

This protocol provides a detailed, step-by-step guide for employing TurboID to identify proximal interacting proteins of a target protein in the model plant *Nicotiana benthamiana*. This species is widely adopted in plant molecular biology and plant-microbe interaction research due to its susceptibility to Agrobacterium-mediated transient transformation, genetic tractability, and short experimental timelines. This workflow covers vector construction, transient transformation of *N. benthamiana* leaves, *in planta* biotin labeling, protein purification, and sample preparation for mass spectrometry. We emphasize optimization steps specific to plant materials and the inclusion of critical controls to ensure reliable results.

2 Materials

2.1 Plasmids

The TurboID plasmid [11] was kindly provided by Prof. Yongliang Zhang (China Agricultural University). The pCV-Myc-TruboID-X vector is used for Myc-TurboID-tagged gene expression vector construction (Fig. 1). This expression vector was modified from the pCV-Myc-X vector which was kept in our laboratory.

2.2 Plant Material and Bacterial Strains

1. 3- to 4-week-old *Nicotiana benthamiana* plants are grown in greenhouse at 22 °C with a light cycle of 16 h, a dark of 8 h, and a relative humidity of 65–70% (*see* **Note 1**).
2. *Escherichia coli* strain DH5α.
3. *Agrobacterium tumefaciens* strain GV3101.

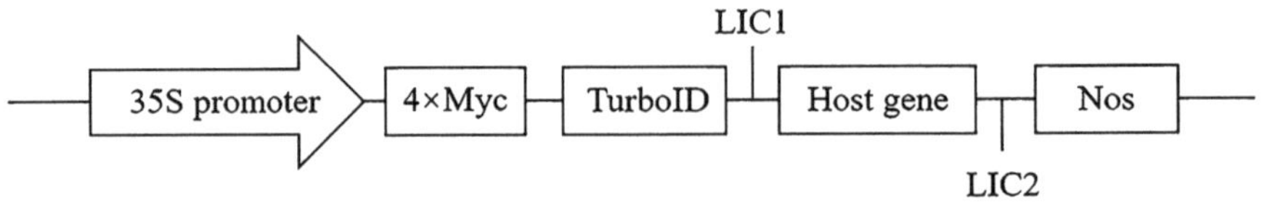

Fig. 1 Schematic diagram of Myc-tagged TurboID dual expression vector

2.3 Preparation of Escherichia/ Agrobacterium Culture

DH5α strain and GV3101 strain are cultured on solid or liquid Luria-Bertani (LB) medium: 1% (W/V) Bacto Tryptone, 0.5% (W/V) yeast extract, 1% (W/V) NaCl, and adjust pH to 7.00 with NaOH. Autoclave at 121 °C for 15 min.

2.4 Agrobacterium Infiltration Buffer

10 mM $MgCl_2$, 10 mM MES (2-(N-morpholino) ethanesulfonic acid) (pH 5.6), 200 μM Acetosyringone (*see* **Note 2**).

2.5 SDS-Polyacrylamide Gel Electrophoresis

1. Separating buffer: 1.5 M Tris-HCl, pH 8.8.
2. Stacking buffer: 1 M Tris-HCl, pH 6.8.
3. 10% (w/v) Sodium dodecyl sulfate (SDS).
4. 10% (w/v) Ammonium persulfate (APS).
5. Thirty percent acrylamide/bis solution (29:1).
6. N, N, N′, N′-Tetramethylethylenediamine.
7. Running buffer (10×): Dissolve 30.3 g Tris-base, 151.1 g glycine, 10 g SDS in 1 L sterile water.
8. Blue Plus® V Protein Marker (10–190 kDa).
9. SDS-PAGE loading buffer (5×): 0.25 M Tris-Cl (pH 6.8), 10% (w/v) SDS, 50% (v/v) glycerol, 5% β-mercaptoethanol, 0.5% bromophenol blue.

2.6 Western Blotting

1. Transfer buffer: 25 mM Tris-base, 192 mM glycine, 0.1% (w/v) SDS, 20% (v/v) Methanol, pH 8.3.
2. Immobilon®-P PVDF Membrane 0.45 μm.
3. Mini Trans-Blot Cell.
4. Tris-buffered saline with Tween (TBS-T) (10×): 1.37 M NaCl, 250 mM Tris–Cl, pH 7.5, 1% (v/v) Tween-20.
5. Blocking solution: 5% (w/v) non-fat milk powder in 1 × TBS.
6. Primary antibody: Anti-c-Myc Mouse Monoclonal Antibody.
7. Secondary antibody: ProteinFind® Goat Anti-Mouse IgG (H + L), HRP Conjugate.

2.7 Critical Reagents and Consumables for TurboID Sample Enrichment Experiments

1. cOmplete™ Protease Inhibitor Cocktail.
2. Zeba™ Spin Desalting Column.
3. Pierce™ BCA Protein Assay Kit.
4. Streptavidin-coated magnetic beads.
5. Biotin immersion solution: 200 μM biotin, 10 mM $MgCl_2$.
6. RIPA lysis buffer: 50 mM Tris–HCl, 500 mM NaCl, 1 mM EDTA, 1% NP40(v/v), 0.1%SDS(w/v), 0.5% sodium deoxycholate, 1×cOmplete™ Protease Inhibitor Cocktail.
7. Washing buffer 1: 2% SDS(w/v).

8. Washing buffer 2: 50 mM HEPES (pH 7.5), 500 mM NaCl, 1 mM EDTA, 0.1% deoxycholic, 1% Triton X-100.
9. Washing buffer 3: 10 mM Tris–HCl (pH 7.5), 250 mM LiCl, 1 mM EDTA, 0.1% deoxycholic, 1% NP40(v/v).

3 Methods

3.1 Construction of the Myc-TurboID-Tagged Gene via Ligation Independent Cloning (LIC) Method

1. Design specific primers containing LIC adapters according to the corresponding CDS of the gene (LIC adapter F: 5′--CgACgACAAgACCgTcacc-3′/LIC adapter R: 5′--gAggAgAagAgCCgTcg-3′) (*see* **Note 3**).
2. Select High-fidelity enzymes of PrimeSTAR series from TAKARA for PCR. 50 μL reaction system consisted of 25 μL of PrimeSTAR® GXL Premix (2×), 1.5 μL of specific upstream primer F, 1.5 μL of specific downstream primer R, 2.0 μL of template cDNA, and 20 μL of ddH_2O. The reaction conditions are 98 °C, 2 min pre-denaturation; (98 °C, 15 s; 55 °C, 20 s; 68 °C, 2 kb/ min)/cycle; 68 °C, 8 min final extension, and generally 30 ~ 35 cycles were performed.
3. Digest the available prokaryotic expression vector pCV-Myc-TruboID-X plasmid with Thermo Scientific™ FastDigest ApaI (*see* **Note 4**).
4. Mix 6 μL of 5× DNA loading buffer with the PCR or digested vector product. Load the samples onto a 1% agarose gel. Run the gel at 200 V for 15 min. Visualize the gel under UV light and compare the band sizes to the DNA marker.
5. Under UV light, excise the gel fragment containing the correctly sized band using a scalpel blade. Weigh the gel slice and proceed with gel recovery. Purify and recover the target DNA fragment by following the protocol of the OMEGA gel extraction kit.
6. Combine 5 μL of the recovered product with 1 μL of 10× buffer, 0.5 μL of 100 nM dATP/dTTP, 0.2 μL of T4 DNA Polymerase, and 3.3 μL of ddH_2O in a reaction tube and mix thoroughly. Incubate the mixture in a PCR machine at 37 °C for 20 min, followed by 75 °C for 20 min.
7. Take 2 μL each of the gene plus A and the vector plus T, mix them and perform the reaction in the PCR instrument, the reaction process is as follows: 70 °C, 20 min; 22 °C, 20 min.
8. Transform all ligation products into *E. coli* DH5α competent cells. Plate the transformation mixture onto LB solid plates containing Kanamycin-resistant. Incubate the plates overnight at 37 °C.
9. Pick single colonies for PCR verification and select positive bacteria for sequencing. Positive clones are defined as those with correct sequencing results.

10. Pick a positive mono-colony, expand it in liquid culture, and then extract the plasmids according to the instructions of OMEGA Plasmid Mini Kit I to obtain Myc-TurboID-tagged gene plasmids.

3.2 Verification of Protein Expression via Western Blot After Transient Overexpression

1. Electroporate Myc-TurboID-tagged plasmids into Agrobacterium tumefaciens GV3101, plate on LB agar with kanamycin and rifampicin, invert, and incubate at 28 °C for 48 h.
2. Pick single colonies for PCR verification to select positive clones, and then add them to LB liquid medium containing corresponding antibiotics and incubate overnight at 200 rpm in a constant temperature incubator at 28 °C.
3. Centrifuge the cells at 5000 rpm for 5 min at room temperature to collect Agrobacterium cells.
4. At the end of centrifugation, decant the supernatant and resuspend the precipitated cells with agrobacterium infiltration buffer. Then, repeat **steps 3** and **4** twice.
5. Resuspend agrobacterium cells containing Myc-TurboID-tagged constructs in infiltration buffer to a final OD_{600} of 0.5 (use an 1800 bp β-glucuronidase (GUS) fragment as a control).
6. Infiltrate Agrobacterium cells harboring Myc-TurboID-tagged constructs into *N. benthamiana* leaves using a 1-mL syringe.
7. At 36 h post-infiltration, collect leaf samples for Western blot analysis. As shown in Fig. 2, both Myc-TurboID-tagged gene *X* and GUS^{1800} control are detectable using an anti-Myc antibody.

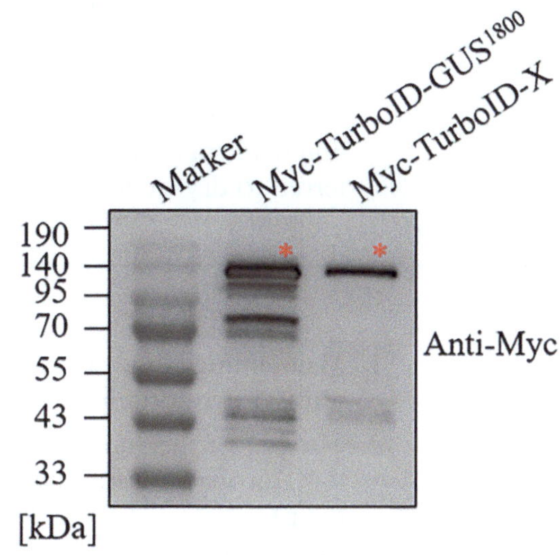

Fig. 2 Verify protein expression via Western blot. Red asterisks point to the expected migration of the fusion proteins

3.3 TurboID Sample Preparation for Mass Spectrometry Analysis

1. Select 3- to 4-week-old *N. benthamiana* plants for each Myc-TurboID-tagged vector. Infiltrate four leaves per plant with *Agrobacterium* suspension ($OD_{600} = 0.5$), using GUS^{1800} as a control.
2. After 36 h, inject the biotin immersion solution into the same leaf (*see* **Note 5**).
3. After completing adjacent labeling through 12 h of dark incubation, harvest the treated leaves, rapidly freeze them in liquid nitrogen, and grind them into powder (90 Hz, 90 sec; *see* **Note 6**).
4. Add 1 mL of RIPA lysis buffer to 700 mg leaf powder to extract the proteins. After vortexing, immediately centrifuge the tube at 16,500 g for 10 min, then remove excess biotin from the lysate by passing the supernatant through a Zeba™ spin desalting column. Next, withdraw 100 μL of lysate from the desalted extract and estimate protein concentration using the Pierce™ BCA protein assay kit (*see* **Note 7**).
5. To enrich biotinylated proteins from protein extracts, take 20 μL of streptavidin-coated magnetic beads and wash twice with 1 mL of RIPA lysis buffer.
6. Incubate the desalted lysate containing ~6 mg of protein overnight with equilibrated magnetic beads on a rotator at 4 °C.
7. Wash once with 1 mL of washing buffer 1; wash once with washing buffer 2; wash once with washing buffer 3 (*see* **Note 8**).
8. To completely remove residual detergent, wash the beads twice in 50 mM Tris–HCl buffer (pH 7.5), followed by six washes in 50 mM NH_4HCO_3 buffer (pH 8.0).
9. Finally, resuspend the beads in 1 mL of 50 mM NH_4HCO_3 buffer. To confirm successful enrichment of the biotinylated protein, extract 10% of the suspension for Western blot analysis. The remaining magnetic beads should be rapidly frozen in liquid nitrogen and stored at −80 °C or immediately transferred to dry ice for LC-MS/MS analysis.

4 Notes

1. Maintaining good watering habits to avoid drought stress of water deprivation in plants before inoculation.
2. The pH of the Agrobacterium infiltration buffer is important for Agrobacterium infection, so it is important to adjust the pH.
3. Attach LIC adapters to the 5′ ends of upstream and downstream primers.

4. The reaction system and timing for Fast Digest ApaI should be performed according to the instructions.
5. Recombinant proteins typically start to express at 36 h post-agroinfiltration.
6. Avoidance of light is necessary to protect the chemical stability of biotin and the biotin-AMP intermediate, thereby ensuring labeling efficiency.
7. Protein concentration measurement is performed to ensure sufficient sample volume for subsequent experiments.
8. The pH of the three washing buffer solutions can affect the experimental results, so they should be freshly prepared before use.

Acknowledgments

The authors are indebted to Prof. Yongliang Zhang (China Agricultural University) for the TurboID plasmid. This research was funded by grants from National Natural Science Foundation of China (32370144) to G.W., Chinese Agriculture Research System of MOF and MARA (CARS-24-C-04) to F.Y., and the K. C. Wong Magna Fund in Ningbo University.

References

1. Fields S, Song O (1989) A novel genetic system to detect protein-protein interactions. Nature 340:245–246
2. Gingras AC, Gstaiger M, Raught B, Aebersold R (2007) Analysis of protein complexes using mass spectrometry. Nat Rev Mol Cell Biol 8: 645–654
3. Kim DI, Roux KJ (2016) Filling the void: proximity-based labeling of proteins in living cells. Trends Cell Biol 26:804–817
4. Rhee HW, Zou P, Udeshi ND, Martell JD, Mootha VK, Carr SA, Ting AY (2013) Proteomic mapping of mitochondria in living cells via spatially restricted enzymatic tagging. Science 339:1328–1331
5. Choi-Rhee E, Schulman H, Cronan JE (2004) Promiscuous protein biotinylation by *Escherichia coli* biotin protein ligase. Protein Sci 13: 3043–3050
6. Roux KJ, Kim DI, Raida M, Burke B (2012) A promiscuous biotin ligase fusion protein identifies proximal and interacting proteins in mammalian cells. J Cell Biol 196:801–810
7. Roux KJ, Kim DI, Burke B, May DG (2018) BioID: a screen for protein-protein interactions. Curr Protoc Protein Sci 74: 19.23.1–19.23.14
8. Branon TC, Bosch JA, Sanchez AD, Udeshi ND, Svinkina T, Carr SA, Feldman JL, Perrimon N, Ting AY (2018) Efficient proximity labeling in living cells and organisms with TurboID. Nat Biotechnol 36:880–887
9. Cho KF, Branon TC, Udeshi ND, Myers SA, Carr SA, Ting AY (2020) Proximity labeling in mammalian cells with TurboID and split-TurboID. Nat Protoc 15:3971–3999
10. Cho KF, Branon TC, Rajeev S, Svinkina T, Udeshi ND, Thoudam T, Kwak C, Rhee HW, Lee IK, Carr SA, Ting AY (2020) Split-TurboID enables contact-dependent proximity labeling in cells. Proc Natl Acad Sci USA 117: 12143–12154
11. Zhang YL, Song GY, Lal NK et al (2019) TurboID-based proximity labeling reveals that UBR7 is a regulator of N NLR immune receptor-mediated immunity. Nat Commun 10:3252

Chapter 8

Protocol for Exogenous Jasmonate Treatment and Quantitative Assessment of Plant Virus Infection

Kelei Han, Chao Ma, and Dankan Yan

Abstract

Plant hormones play a crucial role in mediating antiviral defenses, where different hormones exert diverse effects depending on the specific plant-virus interaction. To conduct a comprehensive assessment of the impact of phytohormones on viral resistance, this chapter proposes a detailed methodology. It takes the role of jasmonic acid (JA) in the tri-partite interactions among *Oryza sativa* (rice, the host plant), rice stripe virus (RSV), and associated vectors (small brown planthopper, SBPH) as an illustration. The experimental protocol encompasses the acquisition of RSV by SBPH, the establishment of a highly viruliferous population, virus inoculation, exogenous hormone application, and quantitative assessment of disease symptoms. This methodology can be adjusted to investigate the function of phytohormone within diverse plant-virus pathosystems.

Key words Plant hormone, Plant–virus interaction, Rice stripe virus, Disease assessment

1 Introduction

Plant viruses pose a significant threat to global agricultural production and food security [1]. These viruses infect a wide range of crops, including rice, tomatoes, and potatoes, leading to symptoms such as mosaics, yellowing, necrosis, stunting, and malformation [2]. About 70–80% of these viruses rely on insect vectors for transmission [3]. Controlling plant viruses is challenging because they are obligate parasites, and once a plant is infected, no chemical agents can directly eliminate them. Current control strategies primarily depend on chemical insecticides to manage the vector, but these pose environmental risks and may lead to insecticide resistance [1, 4]. Breeding resistant varieties offers a more sustainable solution, but reliance on major resistance genes (e.g., the *L* gene for tobamoviruses) carries the risk of resistance breakdown against emerging viral strains [1, 5]. Consequently, it is imperative to investigate alternative strategies, including induced resistance.

Guanwei Wu and Xiaofei Cheng (eds.), *Plant Antiviral Resistance: Methods and Protocols*, Methods in Molecular Biology, vol. 3039,
https://doi.org/10.1007/978-1-0716-5300-5_8,

Plant hormones play important roles in regulating developmental processes and signaling networks involved in plant responses to a wide range of biotic and abiotic stresses [6, 7]. Significant progress has been made in understanding the role of phytohormones in plant responses to viral infection. For instance, pretreatment of plants with salicylic acid (SA) can inhibit the infection of potyviruses, tobacco mosaic virus (TMV), and cucumber mosaic virus (CMV). Conversely, mutations in the SA signaling pathway impair the plant's defense mechanism against viral infections [8–10]. Previous studies have also demonstrated that exogenous application of jasmonic acid (JA) can improve the resistance of plants to various plant viruses, such as turnip mosaic virus (TuMV), rice stripe virus (RSV), and rice black-streaked dwarf virus (RBSDV) [11–13].

To assess the impacts of application of phytohormone(s) on plant antiviral resistance, a standardized protocol was established with RSV-vector-rice serving as a model. RSV causes rice stripe disease and is transmitted by the small brown planthopper (*Laodelphax striatellus*, SBPH) in a persistent, circulative, and propagative manner [14]. The methodology involves the acquisition of RSV by SBPH, the establishment of a highly viruliferous population, virus inoculation via viruliferous SBPH, exogenous hormone application, and disease assessment.

2 Materials

2.1 Biological Materials

1. Rice seeds (Oryza sativa *ssp. japonica* cv. Nipponbare).
2. Virus-free small brown planthoppers (*Laodelphax striatellus*).
3. RSV-infected rice plants as virus source (collected at Anhui province).

2.2 Chemical

1. Ethanol.
2. Tween-20.
3. Primary Polyclonal antibody against RSV coat protein from Rabbit.
4. Goat anti-Rabbit IgG (H + L) Secondary Antibody, AP.
5. Nitrocellulose Membranes, 0.45 μm).
6. BCIP/NBT Chromogen Kit.
7. NON-Fat Powdered Milk.

2.3 Equipment

1. Plant growth chamber with controlled temperature, humidity, and light conditions.
2. 1 L beakers.
3. Laboratory pH meter.

4. Laboratory Magnetic Stirrer.
5. Platform Shaker.
6. 50 mL spray applicator for hormone treatment.

2.4 Buffers

1. 100 mM Methyl jasmonate (MeJA) stock solution: Dissolve 224.3 mg MeJA in ethanol to a final volume of 10 mL.
2. Carbonate coating solution: Dissolve 1.59 g Na_2CO_3, 2.93 g $NaHCO_3$ in distilled water to a final volume of 1 L. Adjust the pH to 9.6.
3. 10× phosphate-buffered saline (PBS): Dissolve 16 g NaCl, 4 g KCl, 4 g KH_2PO_4, and 6 g $Na_2HPO_4{\cdot}12H_2O$ in distilled water to a final volume of 1 L. Adjust the pH to 7.5.
4. PBS-T: 10× PBS diluted by 1:9 volume ratio with distilled water, added 0.1% Tween 20.
5. Blocking solution: (5% (w/v) skimmed milk powder in PBS-T.

3 Methods

3.1 Preparation of Highly Viruliferous Small Brown Planthoppers

1. Maintain virus-free small brown planthoppers on healthy rice plants in beakers at 25 ± 2 °C.
2. Transfer 100–150 2nd to 3rd-instar nymphs to RSV-infected rice plants for a 72-hour acquisition feeding period for virus acquisition (*see* **Note1**).
3. Transfer a pair of male and female SBPH to healthy rice seedlings within a rearing cup (approximately 30 cups) and waiting for oviposition (Fig. 1).
4. After oviposition, retrieve, number, and test female insects for RSV by dot immunobinding assay (DIBA) [15] using anti-RSV coat protein (CP) antibodies.
5. Rear eggs laid by female planthoppers carrying RSV to establish a population of viruliferous *L. striatellus*.
6. Assess the proportion of viruliferous planthoppers via DIBA (Fig. 2) to acquire an adequate quantity (≥500 planthoppers) of highly viruliferous SBPH populations for RSV inoculation (*see* **Notes 2** and **3**).

3.2 Plant Growth and Virus Inoculation

1. Immerse approximately 250 rice seeds at a temperature of 25 °C ± 2 °C for a duration of two days (with a daily light exposure exceeding 12 h), followed by a 2-day germination period.
2. Transplant 35 pre-germinated rice seedlings into 1-L plastic beakers. Cover the beakers with a sheet of nylon mesh, and cultivate the plants in a controlled environment chamber

Fig. 1 Rearing setup for small brown planthopper pairs. Individual female-male pairs were maintained in separate beakers

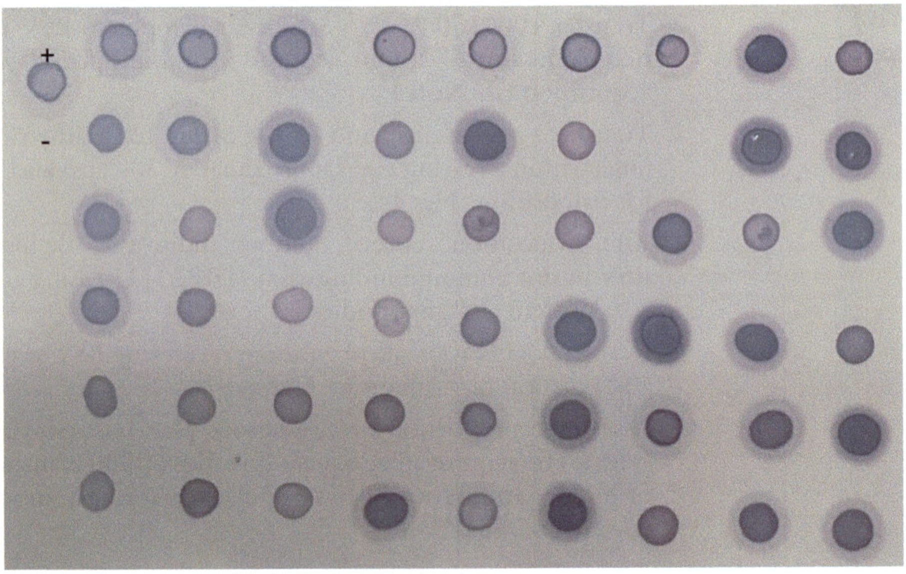

Fig. 2 Proportion of viruliferous small brown planthoppers determined by dot immunobinding assay (DIBA). +: Positive control; –: Negative control

maintained at 25 ± 2 °C, with a relative humidity ranging from 50% to 70% and a photoperiod of 14 h of light and 10 h of darkness (*see* **Note 4**).

3. Use 10-day-old rice seedlings at the 1.5 leaf stage for RSV inoculation.
4. Place two viruliferous small brown planthoppers (SBPHs) on each test rice seedling (with a total of 70 SBPHs per beaker). Provide a 48-hour inoculation access period for virus transmission. During this period, gently agitate SBPHs at 6- to 8-hour intervals to ensure uniform distribution, feeding, and efficient RSV inoculation.

3.3 MeJA Treatment

1. Dilute MeJA stock solution to 100 μM working concentration in ddH_2O containing 0.05% Tween-20 as surfactant. The same concentration of ethanol (0.1%) and surfactant (Tween-20, 0.05%) in ddH_2O was prepared as control buffer (CK) (*see* **Note 5**).
2. Application: 6–8 h after removing the viruliferous SBPH, apply the 100 μM MeJA solution to rice seedlings using a 50 mL sprayer until runoff (approximately 6 ml per beaker). Apply control buffer (0.1% ethanol) to plants using the same method (*see* **Note 6**).
3. Two days after MeJA treatment, transplant rice seedlings to greenhouse or field (Fig. 3; maintain temperature at 15–30 °C) (*see* **Note 7**).

Fig. 3 Transplantation of rice stripe virus-inoculated rice seedlings to field conditions

Fig. 4 Typical symptoms of rice stripe disease across severity grades. (N) No symptoms; (I) Mild chlorotic stripes without leaf curling; (II) Moderate stripes with slight or no leaf curling; (III) Severe stripes with limited necrosis and leaf curling; (IV) Severe stunting, wilting, or plant death

3.4 Disease Assessment and Data Analysis

1. Investigate the disease incidence and disease severity about twenty days after transplanting (approximately 25 days post-inoculation; *see* **Notes 8** and **9**).
2. The disease severity grades of rice stripe disease and their corresponding symptom descriptions are shown in Fig. 4.
3. Based on the descriptions of the disease symptoms (Fig. 4), investigate each plant individually and record the disease severity grading.
4. Calculate the disease index (DI) using the following formula: Disease index = Σ(Disease grade $\times$ Number of plants in that grade) / (4 $\times$ Total plants) $\times$ 100.
5. Employ the t-test statistical approach to conduct an analysis of the Disease Index between the two groups. Statistical significance of differences is considered when $p \leq 0.05$.

3.5 Representative Results

The data presented in the statistical table and the bar chart depicting the percentage of disease severity consistently indicated that MeJA treatment alleviates the severity of rice stripe disease in comparison with the untreated control (Fig. 5a, b). Moreover, the disease index (DI) of rice stripe disease in MeJA-treated rice plants is also significantly lower than that in the control (CK)-treated rice plants (Fig. 5c). Collectively, these results imply that exogenous MeJA treatment confer resistance against rice stripe virus.

A

Treat	Repeat	Total plants	Disease Severity					DI
			N	I	II	III	IV	
MeJA	Rep1	25	8	4	7	3	3	39.00
	Rep2	31	9	8	5	5	4	39.52
	Rep3	28	7	6	7	4	4	42.86
CK (0.1% ethanol)	Rep1	32	5	3	7	9	8	59.38
	Rep2	30	6	4	7	7	6	52.50
	Rep3	26	5	4	6	4	7	53.85

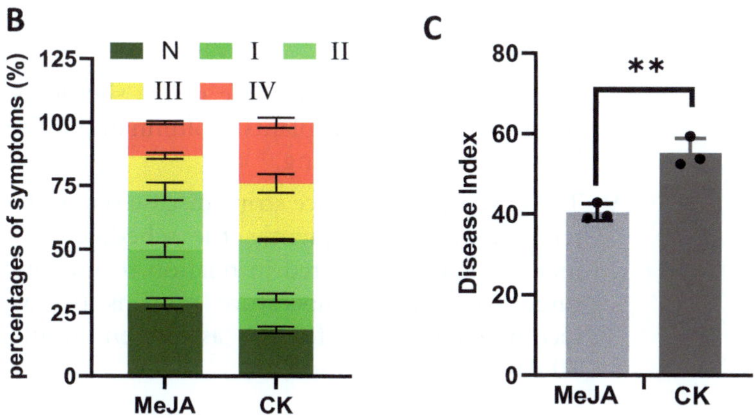

Fig. 5 Effect of exogenous methyl jasmonate (MeJA) on rice stripe virus (RSV) resistance. (**a**) Disease severity classification, (**b**) Distribution of symptom grades in RSV-infected plants treated with 100 μM MeJA or 0.1% ethanol (control), (**c**) Statistical analysis of disease index (DI)

4 Notes

1. RSV-infected rice plants were confirmed by RT-PCR with primers RSV-CP probe F/R [12] for detection of RSV *CP* gene.
2. Typically, a minimum of 50 small brown planthoppers (SBPHs) in the 2nd–3rd instar nymph stage were sampled using the DIAB method to ascertain the proportion of viruliferous planthoppers.
3. When the proportion of viruliferous planthoppers carrying RSV exceeds 90%, this population can be employed in subsequent RSV inoculation experiments. In cases where the proportion is below 90%, an additional round of screening should be conducted. Typically, an RSV-carrying rate of over 90% can be attained within a maximum of three screening rounds.

4. A minimum of six beakers in total are required (three beakers for the application of methyl jasmonate (MeJA) and the other three beakers for the application of the control buffer).
5. Solution preparation: Freshly prepare MeJA solutions before each application as jasmonates can degrade over time. Protect solutions from light and use appropriate solvents to ensure proper solubility.
6. Critical timing: The timing of MeJA application after virus inoculation is critical. Application at 6–8 h post-inoculation corresponds to early stages of viral establishment, potentially interfering with initial replication events.
7. For each treatment, a minimum of three replicate plots must be established, with approximately 30 rice seedlings transplanted per plot.
8. The number of surviving rice plants should be counted and recorded in the "Total plants" column of the Table in Fig. 5a at 7 days after transplanting.
9. The development of rice stripe disease symptoms is affected by temperature. Consequently, the schedule for disease assessment should be adjusted in relation to the ambient temperature. When the temperature remains below 25 °C, it is recommended to defer the assessment until approximately 30 days after inoculation.

Acknowledgments

This work was financially supported by the National Natural Science Foundation of China (32300150) and the Youth Development Fund from Anhui Academy of Agricultural Sciences (QNYC-202506, QNYC-202507).

References

1. Tatineni S, Hein G (2023) Plant viruses of agricultural importance: current and future perspectives of virus disease management strategies. Phytopathology 113(2):117–141
2. Gao M, Lozano-Durán R (2025) Symptom development in plant viral diseases: what, how, and why? Annu Rev Phytopathol 63(1):431–450
3. Dietzgen RG, Mann KS, Johnson KN (2016) Plant virus-insect vector interactions: current and potential future research directions. Viruses 8(11):303
4. van den Berg H, da Silva Bezerra HS, Al-Eryani S et al (2021) Recent trends in global insecticide use for disease vector control and potential implications for resistance management. Sci Rep 11(1):23867
5. Tomita R, Sekine KT, Mizumoto H et al (2011) Genetic basis for the hierarchical interaction between *Tobamovirus* spp. and *L* resistance gene alleles from different pepper species. Mol Plant-Microbe Interact 24(1):108–117
6. Alazem M, Lin NS (2015) Roles of plant hormones in the regulation of host-virus interactions. Mol Plant Pathol 16(5):529–540
7. Carr JP, Murphy AM, Tungadi T et al (2019) Plant defense signals: players and pawns in

plant-virus-vector interactions. Plant Sci 279: 87–95

8. Lee WS, Fu SF, Li Z, Murphy AM et al (2016) Salicylic acid treatment, and expression of an RNA-dependent RNA polymerase 1 transgene inhibit lethal symptoms, and meristem invasion during tobacco mosaic virus infection in *Nicotiana benthamiana*. BMC Plant Biol 16:15
9. Baebler S, Witek K, Petek M, Stare K et al (2014) Salicylic acid is an indispensable component of the *Ny-1* resistance gene-mediated response against potato virus Y infection in potato. J Exp Bot 65:1095–1109
10. Zhou T, Murphy AM, Lewsey MG et al (2014) Domains of the cucumber mosaic virus 2b silencing suppressor protein affecting inhibition of salicylic acid-induced resistance and priming of salicylic acid accumulation during infection. J Gen Virol 95:1408–1413
11. Ji M, Zhao J, Han K et al (2021) Turnip mosaic virus P1 suppresses JA biosynthesis by degrading cpSRP54 that delivers AOCs onto the thylakoid membrane to facilitate viral infection. PLoS Pathog 17:e1010108
12. Han K, Huang H, Zheng H et al (2020) Rice stripe virus coat protein induces the accumulation of jasmonic acid, activating plant defense against the virus while also attracting its vector to feed. Mol Plant Pathol 21(12):1647–1653
13. He Y, Zhang H, Sun Z et al (2017) Jasmonic acid-mediated defense suppresses brassinosteroid-mediated susceptibility to rice black streaked dwarf virus infection in rice. New Phytol 214(1):388–399
14. Xu Y, Fu S, Tao X et al (2021) Rice stripe virus: exploring molecular weapons in the arsenal of a negative-sense RNA virus. Annu Rev Phytopathol 59:351–371
15. Wang G, Zhou Y, Zhou X (2005) Detection of rice stripe virus in *Laodelphax striatellus* by direct dot immunobinding assay. J Zhejiang Univ 31(1):37–40

Chapter 9

Exogenous Melatonin Application Via Foliar Spray and Root Irrigation for Plant Virus Control

Xue Yang, Lingling Yang, and Yan Shi

Abstract

Melatonin, a multifunctional phytohormone, has been demonstrated to enhance plant resistance against various pathogens, including plant viruses. This chapter provides a detailed protocol for the exogenous application of melatonin via two primary methods: foliar spray and root irrigation. The authors describe the preparation of melatonin solutions, optimal concentration ranges, treatment schedules, and practical considerations for maximizing its antiviral efficacy in plants. The methods outlined here are derived from successful applications in controlling plant viruses.

Key words Melatonin, Foliar spray, Root irrigation, Viruses

1 Introduction

Melatonin (N-acetyl-5-methoxytryptamine) is an evolutionarily conserved molecule that regulates various physiological processes in plants [1, 2], including growth, development, and stress responses [3–8]. Beyond its role in abiotic stress tolerance, exogenous melatonin treatment inhibits infection of plants by viral, fungal, and bacterial pathogens. Exogenous melatonin treatment improves resistance to plant pathogens, through activating antioxidant genes, manipulating lignin and gossypol biosynthesis, reactive oxygen and nitrogen species scavenging, cell wall and callose accumulation, hormonal cross-talk, pattern-triggered immunity (PTI) and effector-triggered immunity (ETI) regulation, and pathogenesis-related protein induction [9–12]. In recent years, melatonin has emerged as a promising eco-friendly alternative for inducing broad-spectrum resistance against viral infections [13–16]. Two common delivery methods, foliar spray and root irrigation, have been widely used for melatonin treatment, each with distinct advantages and limitations [13]. Foliar spray allows direct absorption through the leaf surface, while root irrigation facilitates

Guanwei Wu and Xiaofei Cheng (eds.), *Plant Antiviral Resistance: Methods and Protocols*, Methods in Molecular Biology, vol. 3039, https://doi.org/10.1007/978-1-0716-5300-5_9,

systemic distribution via the vascular system. This chapter provides standardized protocols for melatonin application to control plant viruses, optimized through evaluation of concentration, timing, and delivery methods, to facilitate its use in both research and agricultural practice.

2 Materials

2.1 Chemical Reagents

1. Melatonin.
2. Emery powder.
3. RNAiso Plus.
4. Liquid nitrogen.
5. Anhydrous chloroform.
6. Isopropanol.
7. 75% ethanol.
8. RNase-free water.
9. RNase-free DNase I gDNA wiper.
10. HiScript III qRT SuperMix for qPCR.
11. 2 × ChamQ Universal SYBR qPCR Master Mix.
12. Lysis buffer: 100 mM Tris-HCl, pH 8.8; 6% sodium dodecyl sulfate, 2% β-mercaptoethanol.
13. 1 × TBST buffer solution: 8.8 g NaCl, 20 mL 1.0 M Tris–HCl (pH 8), 500 μL Tween-20, make up to 1 L.
14. 15% and 5% SDS-polyacrylamide gel electrophoresis (PAGE) gels.
15. 5 × Loading buffer: 250 mM Tris (pH = 6.8), 10% SDS, 0.5% bromophenol blue, 50% glycerol, 5% β-mercaptoethanol.
16. 5% skimmed milk.
17. Coomassie Brilliant Blue R-250 Staining Solution: 1 g Coomassie Brilliant Blue R-250, 250 mL.
18. Primary and HRP-conjugated secondary antibodies.
19. EasySee Western Blot Kit.
20. Double-distilled water (ddH_2O).

2.2 Equipment

1. Spray bottle (for foliar application).
2. Irrigation can or automated watering system (for root irrigation).
3. Pipettes and volumetric flasks.
4. Grinding instrument.
5. Refrigerated Benchtop centrifuge.

6. Pipettes (20, 100, and 1000 μL).
7. Biosafety cabinet.
8. Nitrocellulose membrane (NC membrane).
9. NanoDrop UV/Vis Spectrophotometer.
10. Gel electrophoresis system.
11. QuantStudio 3 Real-Time Polymerase Chain Reaction (PCR) System.
12. Amersham Imager 680.

3 Methods

3.1 Flowchart of Melatonin Treatment

The experimental design is summarized in Fig. 1. Protocol 1 determines the optimal melatonin application method by treating virus-infected plants via foliar spray or root irrigation (Fig. 1a). Protocol

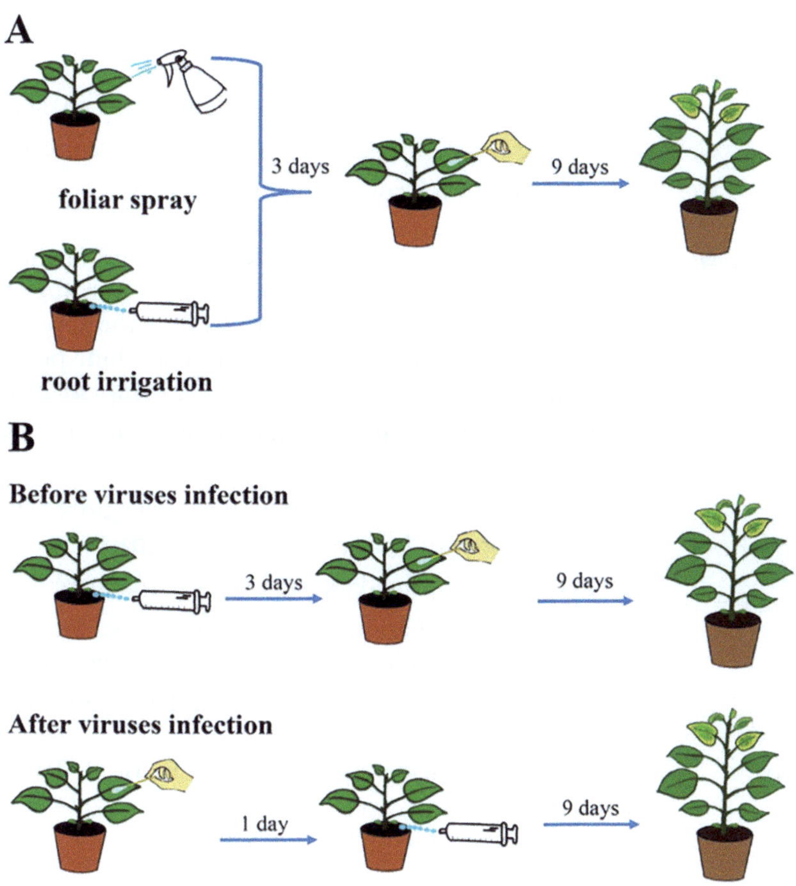

Fig. 1 Flowchart of melatonin treatment strategies in plants. (**a**) Application methods: foliar spray or root irrigation. (**b**) Treatment timing: pre-inoculation (prevention) or post-inoculation (therapy)

2 evaluates melatonin treatment for prevention (pre-inoculation) or therapy (post-inoculation) (Fig. 1b). Viral accumulation and symptom development are subsequently monitored and quantified to assess treatment efficacy.

3.2 Preparation of Melatonin Working Solution

Dilute the melatonin stock solution with ddH_2O to the desired working concentration (e.g., 50 μM, 100 μM; *see* **Notes 1** and **2**). For foliar spray, add 0.01% Tween-20 as a surfactant to improve adhesion (optional).

3.3 Foliar Spray Application

1. Apply melatonin solution evenly to both adaxial and abaxial surfaces of leaves until runoff.
2. Perform spraying in the early morning or late afternoon to avoid rapid evaporation.
3. Repeat daily for 3 days before or after virus inoculation.
4. Include a control group sprayed with 0.01% Tween-20 in ddH_2O.

3.4 Root Irrigation Application (See Note 3)

1. Water plants with melatonin solution until the soil is fully saturated.
2. Apply daily for 3 days before or after virus challenge.
3. Ensure consistent soil moisture to facilitate melatonin uptake.
4. Use ddH_2O for control plants.

3.5 Virus Mechanical Inoculation

1. Collect virus-infected plant leaves and grind thoroughly with 1× PBS buffer in a mortar and pestle. Centrifuge the homogenate at 2000 × g for 10 min. Collect the supernatant as viral inoculum. Gently distribute carborundum powder over the plant leaf surface of three leaves per plant (*see* **Note 4**).
2. Pipette 200 μL of viral inoculum onto the carborundum-dusted area.
3. Gently rub the leaf surface with a sterile gloved finger or a soft applicator with light pressure for 3–5 passes.
4. Keep plants in the dark for 1 h post-treatment.
5. After the incubation period, gently rinse the leaves with tap water to remove residual carborundum and inoculum.
6. Transfer the plants back to the standard plant growth chamber.

3.6 Timing and Concentration

Preventive treatment: Apply melatonin for 3 days before virus inoculation.

Therapeutic treatment: Apply melatonin for 3 days after virus inoculation.

Optimal concentration: 50 μM for both tobacco and cucumber plants (*see* **Note 5**).

3.7 Virus Accumulation Detection

3.7.1 RNA Extraction

1. Pre-chill grinding instrument and 2 mL centrifuge tubes in liquid nitrogen. Grind 50–100 mg virus-infected leaf tissue (fresh or frozen) to fine powder.
2. Transfer powder to tube with pre-cooled RNAiso Plus (1 mL per 50–100 mg tissue) and incubate on ice for 5 min.
3. Centrifuge at 12,000 × g for 5 min at 4 °C. Carefully transfer the supernatant to a new tube, avoiding the pellet.
4. Add 200 μL of chloroform to the supernatant. Securely close the tube and mix vigorously until the solution becomes emulsified and milky in appearance. Let the mixture stand on ice for 10 min.
5. Centrifuge at 12,000 × g for 15 min at 4 °C. Carefully transfer the upper aqueous phase to a new tube without disturbing the interphase.
6. Add 500 μL of isopropanol to the aqueous phase, invert the tube to mix, and incubate at 4 °C for 10 min to precipitate the RNA. Centrifuge at 12,000 × g for 10 min at 4 °C. Discard the supernatant without disturbing the RNA pellet.
7. Wash the pellet by adding 1 mL of 75% ethanol, gently inverting the tube, and then centrifuging at 7500 × g for 5 minutes at 4 °C. Carefully discard the ethanol supernatant.
8. Open the cap of the centrifuge tube, let it dry for a few minutes.
9. Add an appropriate amount of RNase-free water to dissolve the sediment.

3.7.2 Reverse Transcription Quantitative Polymerase Chain Reaction

1. For the genomic DNA removal step, assemble the reaction mixture in an RNase-free microcentrifuge tube by combining 1 pg to 1 μg of total RNA template with 4 μL of 4 × gDNA wiper mix, then adjust the total volume to 16 μL using RNase-free ddH_2O. After gently mixing the components by pipetting, incubate the mixture at 42 °C for 2 min (*see* **Note 6**).
2. Add 4 μL of 5 × HiScript III qRT SuperMix directly to the reaction tube from **step 1** (total volume becomes 20 μL). After gently mixing the components by pipetting, incubate the mixture at 37 °C for 15 min, followed by enzyme inactivation at 85 °C for 5 seconds (*see* **Note 6**).
3. The cDNA product can be utilized directly in qPCR applications or stored at −20 °C for a maximum of 6 months.
4. Run qPCR using 2 × ChamQ Universal SYBR Master Mix with virus-specific primers. Program: 95 °C 30 s pre-denaturation; 40 cycles of 95 °C 10 s, 56 °C 30 s, 72 °C 30 s. Analyze data using 2^-ΔΔCT method (*see* **Notes** 7 and **8**).

3.7.3 Western Blotting

1. Extract 0.1 g crushed plant tissue in 400 μL lysis buffer and incubate at 4 °C for 30 min.
2. Centrifuge at 12,000 rpm, 4 °C, 10 min; collect supernatant. Mix 20 μL sample with 5 μL 5 × loading buffer; boil 10 min, then ice 2–5 min.
3. Briefly centrifuge to collect the contents at the bottom of the tubes.
4. Run NC at 80–100 V until samples enter 15% separating gel, then increase to 130–140 V until dye front reaches bottom.
5. Transfer protein to nitrocellulose membrane at 25 V or 300 mA for 20–30 min.
6. Block NC membrane in 5% skim milk on horizontal shaker at room temperature for 1 h in airtight container (*see* **Note 9**).
7. Incubate the membranes with corresponding primary antibodies (1:5000 dilution).
8. Wash membrane 3× with washing buffer, 15 min each.
9. Incubate the membranes with HRP-conjugated secondary antibodies (1:10,000 dilution).
10. Detect the immunoreactive protein bands using the EasySee Western Blot Kit and image with Amersham Imager 680.

4 Notes

1. Melatonin is light-sensitive; prepare solutions under dim light and store in amber bottles.
2. Avoid applying melatonin under high-temperature or strong sunlight conditions to prevent degradation.
3. Root irrigation generally provides more stable and systemic melatonin distribution than foliar spray.
4. Make sure the carborundum powder is spread evenly, and keep the number of friction times and the force consistent.
5. Test melatonin concentrations between 10 and 100 μM for different plant species and virus systems.
6. The 4 × gDNA wiper Mix and 5 × HiScript III qRT SuperMix contain high glycerol concentrations. Briefly centrifuge tubes before use and mix components thoroughly by gentle pipetting.
7. The recommended total RNA input for a 20-μL reverse transcription reaction is ≤1 μg. For low-abundance targets, up to 5 μg total RNA may be used. Higher amounts may affect subsequent qPCR linearity.

8. The cDNA product is suitable for qPCR but not recommended for long-fragment PCR required for cloning.
9. Ensure that the solution can cover the NC membrane.

Acknowledgments

This work was funded by the Program for Innovative Research Team (in Science and Technology) in University of Henan Province (25IRTSTHN031) and the Zhongyuan Science and Technology Innovation Youth Talents Program (30603407).

References

1. Dubbels R, Reiter RJ, Klenke E et al (1995) Melatonin in edible plants identified by radioimmunoassay and by high performance liquid chromatography-mass spectrometry. J Pineal Res 18:28–31
2. Hattori A, Migitaka H, Iigo M et al (1995) Identification of melatonin in plants and its effects on plasma melatonin levels and binding to melatonin receptors in vertebrates. Biochem Mol Biol Int 35:627–634
3. Wang Y, Reiter RJ, Chan Z (2018) Phytomelatonin: a universal abiotic stress regulator. J Exp Bot 69:963–974
4. Li H, Guo Y, Lan Z et al (2021) Methyl jasmonate mediates melatonin-induced cold tolerance of grafted watermelon plants. Hortic Res 8:57
5. Zhan Y, Wu T, Zhao X et al (2021) Comparative physiological and full-length transcriptome analyses reveal the molecular mechanism of melatonin-mediated salt tolerance in okra (*Abelmoschus esculentus* L.). BMC Plant Biol 21:180
6. Antoniou C, Chatzimichail G, Xenofontos R et al (2017) Melatonin systemically ameliorates drought stress-induced damage in *Medicago sativa* plants by modulating nitro-oxidative homeostasis and proline metabolism. J Pineal Res 62(4)
7. Ren J, Yang X, Zhang N et al (2022) Melatonin alleviates aluminum-induced growth inhibition by modulating carbon and nitrogen metabolism, and reestablishing redox homeostasis in *Zea mays* L. J Hazard Mater 423:127159
8. Byeon Y, Lee HY, Hwang OJ et al (2015) Coordinated regulation of melatonin synthesis and degradation genes in rice leaves in response to cadmium treatment. J Pineal Res 58(4): 470–478
9. Li C, He Q, Zhang F et al (2019) Melatonin enhances cotton immunity to Verticillium wilt via manipulating lignin and gossypol biosynthesis. Plant J 100:784–800
10. Yin L, Wang P, Li M et al (2013) Exogenous melatonin improves Malus resistance to Marssonina apple blotch. J Pineal Res 54(4): 426–434
11. Mandal MK, Suren H, Ward B et al (2018) Differential roles of melatonin in plant-host resistance and pathogen suppression in cucurbits. J Pineal Res 65:e12505
12. Chen L, Wang MR, Li JW et al (2019) Exogenous application of melatonin improves eradication of apple stem grooving virus from the infected in vitro shoots by shoot tip culture. Plant Pathol 68:997–1006
13. Yang LL, Li QL, Han XY et al (2023) A cysteine-rich secretory protein involves in phytohormone melatonin mediated plant resistance to CGMMV. BMC Plant Biol 23:215
14. Zhao L, Chen L, Gu P et al (2019) Exogenous application of melatonin improves plant resistance to virus infection. Plant Pathol 68:1287–1295
15. Lu R, Liu Z, Shao Y et al (2019) Melatonin is responsible for rice resistance to rice stripe virus infection through a nitric oxide-dependent pathway. Virol J 16(1):141
16. Sofy AR, Sofy MR, Hmed AA et al (2021) Molecular characterization of the alfalfa mosaic virus infecting *Solanum melongena* in Egypt and the control of its deleterious effects with melatonin and salicylic acid. Plants 10:459

Chapter 10

Phosphatidic Acid Application to Enhance Plant Virus Resistance

Jiayu Lin, Wei Zhang, Hongtao Zhang, Bingjian Sun, and Hangjun Sun

Abstract

Phosphatidic acid (PA), a key signaling molecule in plant immunity, activates MAPK-mediated and reactive oxygen species (ROS)-mediated defense responses. Exogenous application of PA significantly reduces Potato virus Y (PVY) genomic RNA accumulation and alleviates disease symptoms in *Nicotiana benthamiana*, highlighting its potential as an immune elicitor for crop disease management. This chapter provides detailed protocols for PA solution preparation, MAPK pathway activation analysis (by Western blot and qPCR), and virus quantification to guide PA application for broader plant virus control in crops.

Key words Phosphatidic acid, Plant antiviral resistance, MAPK, WIPK/SIPK, Potato virus Y

1 Introduction

Phosphatidic acid (PA) is the simplest phospholipid molecule and serves as an intermediate in the synthesis of many other phospholipids. Its concentration in plants is relatively low, ranging from approximately 50–150 μM in *Arabidopsis*, making it one of the least abundant phospholipids in plant cells [1]. However, beyond its role as a structural component of membranes, PA is particularly noteworthy for its function as a signaling molecule, playing a crucial role in regulating plant growth, development [2], disease resistance [3], and stress responses [4].

Upon pathogen invasion, the levels of PA can rapidly increase. Studies have shown that PA levels rise within minutes in plant cells upon recognition of various pathogen-associated molecular patterns (PAMPs), including chitosan oligosaccharides, xylanase, and flg22, as well as in response to pathogen effector proteins such as bacterial AvrRpm1, AvrRpt2, AvrBsT, and fungal AvR4 [5–11]. This indicates that the increase in PA is an early signaling

Jiayu Lin and Wei Zhang contributed equally with all other contributors.

Guanwei Wu and Xiaofei Cheng (eds.), *Plant Antiviral Resistance: Methods and Protocols*, Methods in Molecular Biology, vol. 3039, https://doi.org/10.1007/978-1-0716-5300-5_10,

response common to both pattern-triggered immunity (PTI) and effector-triggered immunity (ETI), similar to the bursts of reactive oxygen species and calcium influx events. The phospholipase D (PLD) and diacylglycerol kinase (DGK) families of enzymes are primarily responsible for the production of PA in plant cells. Evidence suggests that both of these protein families are involved in the "PA burst" during plant responses to pathogen infection [3, 12].

PA can bind to certain target proteins through the negative charge of its phosphate group, thereby regulating their activity. In terms of plant immune responses, it is well established that PA can interact with MPK3/6 kinases and RBOHD, enhancing their enzymatic activity and subsequently activating MAPK-mediated and reactive oxygen species-mediated immune responses [3, 12, 13]. Application of PA has been shown to significantly inhibit infections by viruses, fungi, and bacteria, indicating its potential as an endogenous immune elicitor in plants, with broad applications in crop disease management [3, 12, 13]. Here, this chapter describes a detailed method for enhancing plant resistance to virus infection using PA, including PA solution preparation, MAPK pathway activation assessment, and virus concentration quantification, using potato virus Y (PVY) and *Nicotiana benthamiana* as a case.

2 Materials

2.1 Biological Materials

1. *N. benthamiana* (5-week-old) are cultivated in a growth chamber at 25 °C with a 16-h light/8-h dark cycle.
2. PVY^N strain (GenBank accession number: MG591487.1) is maintained in Molecular Plant Pathology Laboratory, Henan Agricultural University.

2.2 Preparation of Phosphorylation Assay

1. 50 mM PA: dissolve 0.033 g PA in 1 mL of chloroform in a 2-mL centrifuge tube, vortexing thoroughly until achieving complete dissolution.
2. 50 mM Phosphatidylcholine (PC): dissolve 0.039 g PC in 1 mL chloroform.
3. Protein extraction buffer (Beyotime).
4. Phenylmethyl sulfonyl fuoride (PMSF).
5. Loading buffer: 300 mM Tris–HCl (pH 6.8), 0.1% bromophenol blue, 8% sodium dodecylsulfate (SDS), 25% β-mercaptoethanol, 40% glycerol, and then set the volume to 10 mL by ddH_2O.
6. Protein gel preparation kit.
7. Prestained protein marker, 15–180 kDa.

8. SDS-PAGE running buffer: Dissolve 3.03 g Tris/Cl, 14.4 g glycine and 1 g SDS in 1 L ddH_2O.
9. Transfer buffer: Dissolve 5.82 g Tris/Cl, 2.93 g glycine, 200 mL methanol in ddH_2O, and adjust the final volume to 1 L.
10. Electrophoresis and transfer systems.
11. Polyvinylidene fluoride membrane (PVDF).
12. Tris-buffered saline with Tween (TBS-T) (1×): Dissolve 8.80 g NaCl, add 20 mL 1 M Tris/Cl (pH 7.5) and 500 μL Tween 20, fix the volume to 1 L by ddH_2O.
13. Blocking buffer: Dissolve 1 g non-fat dry milk powder in 20 mL TBS-T (1×).
14. Phospho-p44/42 MAPK (Erk1/2) (Thr202/Tyr204) Rabbit mAb (Cell Signaling Technology, Catalog no. 9101).
15. Goat anti-rabbit IgG conjugated to HRP.
16. Color development reagent: Chemiluminescence kit.
17. Amersham imager 680.
18. Coomassie brilliant blue staining solution: Dissolve 0.25 g Coomassie brilliant blue R250, 45 mL methanol, 10 mL glacial acetic acid in ddH_2O, and adjust the final volume to 100 mL.
19. Coomassie brilliant blue decolorization solution: Add 45 mL methanol and 10 mL glacial acetic acid, and adjust the final volume to 100 mL.

2.3 Preparation of Viral Quantification and Symptom Analysis

1. 10 × PBS buffer.
2. Trizol reagent: RNAiso Plus.
3. HiScript III RT SuperMix.
4. RNase Free ddH_2O.
5. Quantitative PCR kit: ChamQ Universal SYBR qPCR Master Mix.
6. Quantitative PCR machine: ABI7500 System.

3 Methods

3.1 Method of Application of PA

1. Dilute 10 μL phosphatidic acid (PA) or phosphatidylcholine (PC) stock solution (50 mM) in 10 mL ddH_2O and mix thoroughly by vortexing to reach a final concentration of 50 μM. Additionally, prepare a 0.1% chloroform dilution by adding 10 μL of chloroform to 10 mL ddH_2O, followed by vigorous vortex mixing.
2. Select uniform and healthy *N. benthamiana* plants. Evenly spray the adaxial (upper) surface of the fourth and fifth leaves

with 50 μM PA solution using a handheld sprayer. Treat two leaves per plant, applying a total of 1 mL solution per leaf through five sequential sprays to ensure uniform droplet distribution.

3. For control treatments, use an equal volume of either 50 μM PC solution or 0.1% chloroform dilution, maintaining the same spraying method, number of applications, and volume as the PA treatment.
4. Collect all treated leaves at 3 h after treatment, immediately freeze in liquid nitrogen, and store at −80 °C.

3.2 Activation of MAPK Pathway by PA

3.2.1 WIPK/SIPK Phosphorylation Assay

1. Grind liquid nitrogen-frozen samples into fine powder, transfer 0.7 g powder to a 2-mL microcentrifuge tube, add 300 μL protein extraction buffer and PMSF to a final concentration of 1 mM, and mix thoroughly to dissolve.
2. Incubate the protein extract on ice for 10 min to ensure complete cooling, then transfer it to a pre-chilled centrifuge and centrifuge at 12,000 rpm for 10 min at 4 °C. Carefully aspirate the supernatant and transfer it to a new 0.2 mL centrifuge tube.
3. Add 80 μL protein extraction buffer and 20 μL loading buffer to a 0.2-mL centrifuge tube, vortex thoroughly to mix, and heat in a 99 °C water bath for 10 min.
4. Prepare a 10% protein gel with a total volume of 5 mL using a protein gel preparation kit according to the manufacturer's instructions, and allow it to solidify at 27 °C for approximately 20 min.
5. After correctly assembling the gel plate, add 1 L SDS-PAGE running buffer and load 10 μL sample into each well of the gel.
6. Connect to the power supply, set a constant voltage of 120 V, and perform electrophoresis at 4 °C until the dye front reaches the bottom of the gel, which takes approximately 2 h.
7. Soak the size of 8.5 × 4 cm PVDF membrane in methanol for 1 min.
8. Prepare 1 L transfer buffer and perform protein transfer at 4 °C with a constant current of 400 mA for approximately 50 min, until complete protein transfer to the membrane is achieved.
9. At the end of the run, carefully remove the membrane and incubate it in 20 mL blocking buffer at room temperature for 1 h.
10. Dilute 10 μL phospho-p44/42 MAPK (Erk1/2) (Thr202/Tyr204) Rabbit monoclonal antibody in 20 mL blocking buffer at a 1:2000 ratio, then incubate the membrane with the primary antibody at 4 °C overnight.

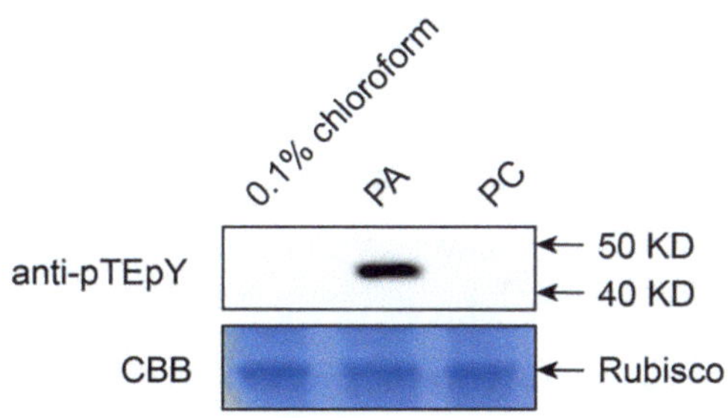

Fig. 1 Phosphatidic acid (PA) stimulates WIPK phosphorylation. *N. benthamiana* leaves were sprayed with 50 μM PA or phosphatidylcholine (PC; control) and harvested after 3 h. Phosphorylated WIPK was detected by Western blotting with a phospho-p44/42 MAPK antibody

11. Wash the membrane three times with 20 mL TBS-T buffer, for 5 min each time.
12. Dilute 4 μL goat anti-rabbit IgG conjugated to HRP (usually 1:5000) in 20 mL blocking buffer and incubate the membrane for 1 h at room temperature.
13. Wash the membrane five times with 20 mL TBS-T buffer, for 5 min each time.
14. Thoroughly mix 100 μL each of color development reagent solutions A and B, apply evenly to the membrane, and image using an Amersham Imager 680.
15. For the staining procedure, incubate the gel in 30 mL coomassie brilliant blue staining solution at room temperature overnight after electrophoresis. Then destain the gel overnight in 30 mL coomassie brilliant blue decolorization solution at room temperature. Once the gel is destained and protein bands become clearly visible, document the results by capturing images using a camera.
16. Compared to the PC treatment, the phosphorylation level of WIPK in *N. benthamiana* are significantly increased at 3 h after treatment with 50 μM PA (Fig. 1).

3.2.2 Detection of the Transcriptional Activation of WRKY8 and HMGR2

1. Extract total RNA using Trizol reagent, then synthesize 10 μL cDNA from 1 μg RNA with HiScript III RT SuperMix according to the manufacturer's instructions. Dilute the synthesized cDNA to 100 μL RNase Free ddH_2O to obtain the template for qPCR.
2. Add the following reagents in 0.2 mL enzyme-free sterile centrifuge tubes: 5 μL ChamQ Universal SYBR qPCR Master Mix, 0.2 μL forward primers (10 μM, Table 1) and reverse primers (10 μM, Table 1), 1 μL template, and 3.6 μL RNase Free ddH_2O, followed by vigorous vortex mixing.

Table 1
Primers used in this protocol

Primer names	Sequence (5′-3′)
NbWRKY8-YF	TCTTTAGCCGTCCAGCCTTACA
NbWRKY8-YR	CACTTTCGCTTTCTCTTTTCCA
NbHMGR2-YF	GTGGCGTGAGAAAATCAGGAA
NbHMGR2-YR	CTGGAAACGAACGACTGGACA
PVY-YF	TGTACTGCGATGCCGATGG
PVY-YR	GGTAGAAGGTTGACCGCTATTATT

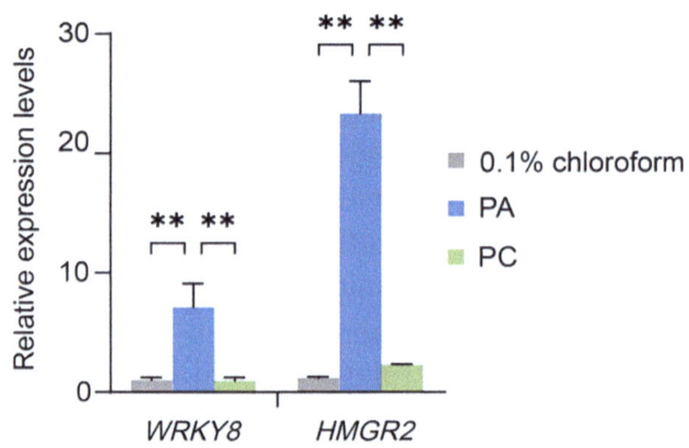

Fig. 2 Phosphatidic acid (PA) upregulates *WRKY8* and *HMGR2* transcription. Following 3 h treatment with 50 μM PA, mRNA levels were quantified by qRT-PCR. Phosphatidylcholine (PC) served as the lipid control. Data represent mean ± SD from three independent replicates; $^{**}P < 0.01$ (Student's t-test)

3. Place centrifuge tubes in an ABI7500 instrument and run qPCR under these conditions: 95 °C for 30 s; 40 cycles of 95 °C for 10 s and 60 °C for 30 s; followed by melt curve analysis at 95 °C for 15 s, 60 °C for 60 s, and 95 °C for 15 s.
4. Use 18 s RNA as an internal control, perform all the experiments in three biological repeats, calculate relative expression levels using the $2^{-\Delta\Delta Ct}$ algorithm, and conduct statistical analysis with Microsoft Excel.
5. The transcriptional levels of *WRKY8* and *HMGR2* are significantly increased at 3 h after spraying with 50 μM PA compared with PC (Fig. 2).

3.3 Asses the Effect of PA Treatment on Plant Virus Infection

1. Here, we use PVY as a case. Collect *N. benthamiana* leaves infected with PVY and place them in a mortar. Gently grind the leaves while adding 5 mL 10 × PBS buffer, then gradually add

sterile water to a final volume to 40 mL. Continue grinding until a homogeneous slurry is formed.

2. Select the third and fourth leaves of similar size from different *N. benthamiana* plants and dust uniformly with 600-grit emery powder. Dip a sterile cotton swab into the inoculation solution and gently rub the emery-dusted leaves. Two to three leaves are inoculated per plant, with each leaf being rubbed 10 times. Incubate inoculated leaves for 2 min, then rinse thoroughly with sterile water to remove residual emery. Transfer plants to normal greenhouse conditions.
3. At 12 h post-inoculation, spray the infected leaves with 50 μM PA or PC. Apply 1 mL solution per leaf per treatment, distributing it evenly in five sprays to ensure full coverage. Repeat the treatment every 48 h.
4. For viral RNA quantification, treat PVY-infected *N. benthamiana* leaves with PA or PC until 3 days post-inoculation (dpi). Collect leaf samples after the final treatment, immediately freeze in liquid nitrogen, and store at −80 °C. Extract total RNA and synthesize cDNA after passing quality control for subsequent qPCR analysis.
5. Use 18 s RNA as an internal control, detect the mRNA contents of PVY by qPCR reaction using PVY-YF (Table 1) and PVY-YR (Table 1) primers following the steps in Subheading 3.2.2.
6. For phenotypic analysis, treat PVY-infected *N. benthamiana* leaves with PA or PC every 48 h until 6 dpi. Observe and compare the symptoms of PVY-infected plants across different treatment groups, capture images of representative symptoms using a camera. As shown in Fig. 3, spray with 50 μM PA significantly inhibits PVY genomic RNA accumulation compared with the PC control group (Fig. 3a). Furthermore, the symptoms observed on the top leaves of PVY-infected plants treated with 50 μM PA are milder than the PC control group at 6 dpi (Fig. 3b).

4 Notes

1. Unless otherwise specified, all reagents and buffers should be prepared using sterile ultrapure water at room temperature and stored according to the temperature conditions recommended in the manufacturer's instructions.
2. PA and PC stock solutions should be stored at −20 °C. It is best to prepare PA and PC working solutions fresh for immediate use, and the number of sprays applied to different leaves should be kept consistent.

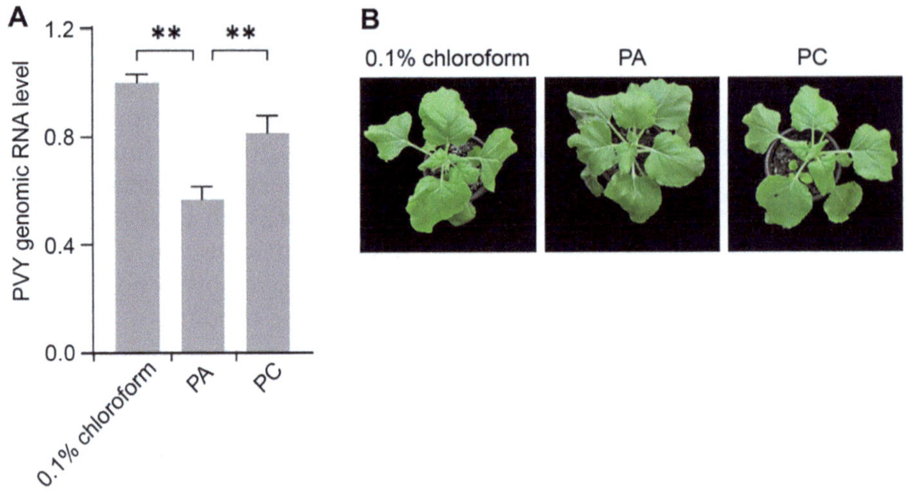

Fig. 3 Phosphatidic acid (PA) exerts antiviral activity against Potato virus Y (PVY). (**a**) Relative PVY genomic RNA levels at 3 days post-inoculation (dpi) in *N. benthamiana* leaves sprayed with 50 μM PA or phosphatidylcholine (PC; negative control). Data represent mean ± SD from three biological replicates. Asterisks denote significant differences (***P* < 0.01, two-tailed Student's t-test). (**b**) Disease symptoms at 6 dpi in plants treated as in (**a**)

3. The concentration range for PA is 50–150 μM. Concentrations that are too high or too low cannot effectively activate the MAPK pathway.
4. When preparing the PA solution, PA powder with a purity greater than 60% should be used. Excessive other lipid components in the PA powder can affect the activation of the MAPK pathway.
5. The timing of PA application is crucial; applying PA five days after virus inoculation has minimal effect on virus control. It is best to apply PA within 24 h of virus inoculation for optimal efficacy.
6. To minimize mechanical damage to the leaves as much as possible, the entire process of virus inoculation should avoid excessive friction forces. The number of friction applications on the inoculated leaves should remain consistent to ensure approximately uniform viral accumulation in each plant.
7. Using the Phospho-p44/42 antibody to detect phosphorylated WIPK/SIPK can sometimes reveal two bands, indicating that both WIPK and SIPK are activated simultaneously. At other times, only one band may be detected, which means that only one of the kinases is activated.
8. It is recommended to incubate with the primary antibody under static conditions at 4 °C overnight. It can effectively reduce the nonspecific binding of antibody and avoid strong back-ground noise during scanning imaging.

9. Replication of experiments: To ensure the reliability of the results, experiments were repeated multiple times. The replication of experiments reduced random errors and strengthened the confidence in the obtained results.

Acknowledgments

This work was supported by the National Natural Science Foundation of China (32572786), the Science and Technology Project of Guizhou Provincial Tobacco Corporation (2025XM03) and the Program for Innovative Research Team in University of Henan Province (25IRTSTHN031).

References

1. Wang X, Devaiah SP, Zhang W et al (2006) Signaling functions of phosphatidic acid. Prog Lipid Res 45:250–278
2. Lin D, Yao H, Jia L et al (2020) Phospholipase D-derived phosphatidic acid promotes root hair development under phosphorus deficiency by suppressing vacuolar degradation of PIN-FORMED2. New Phytol 226:142–155
3. Lin J, Zhao J, Du L et al (2024) Activation of MAPK-mediated immunity by phosphatidic acid in response to positive-strand RNA viruses. Plant Commun 5:100659
4. Zhou Y, Zhou D, Yu W et al (2021) Phosphatidic acid modulates MPK3- and MPK6-mediated hypoxia signaling in *Arabidopsis*. Plant Cell 34:889–909
5. Johansson ON, Fahlberg P, Karimi E et al (2014) Redundancy among phospholipase D isoforms in resistance triggered by recognition of the *Pseudomonas syringae* effector AvrRpm1 in *Arabidopsis thaliana*. Front Plant Sci 5:639
6. Kirik A, Mudgett MB (2009) SOBER1 phospholipase activity suppresses phosphatidic acid accumulation and plant immunity in response to bacterial effector AvrBsT. Proc Natl Acad Sci USA 106:20532–20537
7. Andersson MX, Kourtchenko O, Dangl JL et al (2006) Phospholipase-dependent signalling during the AvrRpm1- and AvrRpt2-induced disease resistance responses in *Arabidopsis thaliana*. Plant J 47(6):947–959
8. de Jong CF, Laxalt AM, Bargmann BO et al (2004) Phosphatidic acid accumulation is an early response in the Cf-4/Avr4 interaction. Plant J 39(1):1–12
9. Yamaguchi T, Minami E, Ueki J et al (2005) Elicitor-induced activation of phospholipases plays an important role for the induction of defense responses in suspension-cultured rice cells. Plant Cell Physiol 46:579–587
10. Yamaguchi T, Minami E, Shibuya N (2003) Activation of phospholipases by N-acetylchitooligosaccharide elicitor in suspension-cultured rice cells mediates reactive oxygen generation. Physiol Plant 118:361–370
11. van der Luit AH, Piatti T, van Doorn A et al (2000) Elicitation of suspension-cultured tomato cells triggers the formation of phosphatidic acid and diacylglycerol pyrophosphate. Plant Physiol 123:1507–1516
12. Kong L, Ma X, Zhang C et al (2024) Dual phosphorylation of DGK5-mediated phosphatidic acid burst regulates ROS in plant immunity. Cell 187(3):609–623
13. Qi F, Li J, Ai Y et al (2024) DGK5β-derived phosphatidic acid regulates ROS production in plant immunity by stabilizing NADPH oxidase. Cell Host Microbe 32:425–440

Chapter 11

Virus Elimination by Thermotherapy Combined with Meristem Culture

Ni Hong and Guoping Wang

Abstract

High-temperature treatment (thermotherapy) inhibits viral RNA synthesis, restricting viral particle movement into the apical meristem and enabling virus elimination when applied for optimized durations. Combined thermotherapy and meristem culture enhances the production of virus-free germplasm. Here, we describe optimized protocols for thermotherapy, meristem culture, and transplantation of virus-free pear plants.

Key words In vitro culture, Heat treatment, Virus detection, Virus-free germplasm

1 Introduction

Utilization of certified virus-free planting material is the most effective strategy for controlling plant viral diseases, enhancing growth vigor, and improving fruit yield and quality. Because viruses are unevenly distributed in plants, meristems are often virus-free and represent ideal tissues for virus elimination [1–3]. However, excision and regeneration of very small meristem tips (~0.25 mm) is technically challenging. Thermotherapy (heat treatment) of in vivo or in vitro plants reduces virus titers and improves virus eradication efficiency from shoot tips [4–7]. Heat stress appears to restrict viral particle movement into the apical meristem by inhibiting viral RNA synthesis [8]. Consistently, studies on in vitro plants demonstrated that thermotherapy reduces virus titer more drastically in shoot tips than in stem bases [8]. Therefore, optimized high-temperature treatments effectively eliminate viruses from apical meristems. Thermotherapy combined with meristem culture [9–13], shoot tip grafting [14], or shoot tip cryotherapy with in vitro culture [7, 15] markedly enhances virus eradication. These approaches are now widely used for virus elimination from fruit trees and other vegetatively propagated plants.

Guanwei Wu and Xiaofei Cheng (eds.), *Plant Antiviral Resistance: Methods and Protocols*, Methods in Molecular Biology, vol. 3039, https://doi.org/10.1007/978-1-0716-5300-5_11,

However, high-temperature treatment reduces plant survival rates, limiting the efficiency of thermotherapy-based virus elimination. For example, sand pear is heat-sensitive: only 56% of in vitro plants survived alternating temperature (34 °C /42 °C) thermotherapy for 50 days [16], and 14.2% survived constant 37 °C thermotherapy for 35 days [9], Therefore, treatment temperature and duration must be optimized for each plant species and target virus, with typical protocols ranging from 36 to 38 °C for 21–40 days [5, 11, 12], or alternating 34 °C/42 °C for 40–-60 days [16]. Shoot-tip size excised from thermotherapy-treated plants also significantly influences virus elimination efficiency [8, 16].

Here, we provide a detailed, highly effective protocol for virus eradication using thermotherapy combined with meristem culture in pear.

2 Materials

2.1 Plant Materials

The pear plants used for thermotherapy can be potted young plants or in vitro cultured plants.

2.2 MS Medium

1. Macro-element stock solution (1 L, 10×): Weigh 19 g KNO_3, 16.5 g NH_4NO_3, 3.7 g $MgSO_4{\cdot}7H_2O$, 1.7 KH_2PO_4 g, 4.4 g $CuCl_2{\cdot}2H_2O$, or 3.4 g $CaCl_2$ into a glass beaker. Dissolve with 900 mL of water. Make up to 1 L with water. Store at 4 °C.
2. Micro-element solution (1 L, 100×): Add 200–300 mL water into a glass beaker. Weigh 0.083 g KI, 0.62 g H_3BO_3, 0.86 g $ZnSO_4{\cdot}7H_2O$, 0.025 g $NaMoO_4{\cdot}2H_2O$, 0.0025 g $CuSO_4{\cdot}5H_2O$, 0.0025 g $CoCl_2{\cdot}6H_2O$, and 2.23 g $MnSO_4{\cdot}4H_2O$. Individually add them into the glass beaker and vortex to dissolve them. Make up to 1 L with water. Place at room temperature to avoid forming pellets. Store at 4 °C.
3. Glycine (100×): Dissolve 0.2 g Glycine in 1 L water.
4. Inositol (100×): Dissolve 10 g Inositol in 1 L water.
5. Fe salt (100×): Dissolve 3.73 g EDTA and 2.78 g FeSO4 in 1 L water in a dark glass bottle. Store at 4 °C (*see* **Note 4**).
6. Vitamin B (VB, 100×): Dissolve 50 mg thiamine hydrochloride (VB_1), 10 mg pyridoxine hydrochloride (VB_6), and 50 mg nicotinic acid (VB_5) in 1 L water.
7. Vitamin C (VC, 100×): Dissolve 0.5 g VC in 1 L water.
8. Phytohormone: 6-Benzylaminopurine (6-BA), 3-indole-butyric acid (IBA),

9. Basic Murashige and Skoog (MS) medium: Mix 100 mL macro-element stock solution, 10 mL micro-element solution, 10 mL glycine, 10 mL VB, 10 mL VC, 10 mL Fe salt, and 10 mL inositol. Make up to 1 L with water.
10. Culture medium: MS medium complimented with 0.2 mg/L IBA and 1.0 mg 6-BA. Make up to 1 L with water. Adjusted pH to 5.8 using 1 mol/L NaOH. Add 30 g sucrose and 7.5 g agar. Autoclave at 121 °C for 15–20 min.
11. Rooting medium: 1/2 MS supplemented with 0.5–2.5 mg/L IBA, 1.5 mg/L1-naphthaleneacetic acid (NAA), 2.0 g/L sucrose and 5.8–6.0 g/L agar (*see* **Note 5**).
12. Transplant medium: Mix commercial nutrient soil, vermiculite, and peat at a volume ratio of 5:2:1.

3 Methods

3.1 Flowchart of Thermotherapy Combined with Meristem Culture

The procedure is schematically illustrated in Fig. 1. The workflow for virus elimination via thermotherapy and meristem culture involves heat-treating in vitro cultures, followed by excision and multiplication of shoot tips on MS medium [17]. Virus-free status is confirmed through reverse transcription polymerase chain reaction (RT-PCR) and dot hybridization screening, after which cultures are rooted, transplanted to pots, and maintained in insect-proof greenhouses before establishing virus-free mother plants in the field.

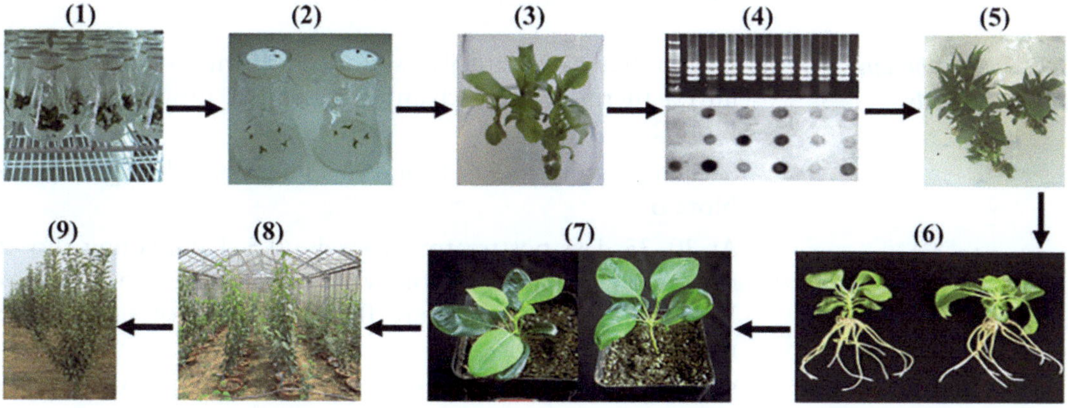

Fig. 1 Schematic workflow for pear virus elimination by thermotherapy combined with meristem culture. (1) In vitro cultures subjected to heat treatment in a growth chamber. (2–3) Shoot tip isolation from heat-treated plants and proliferation on MS medium. (4) Virus detection by multiplex RT-PCR and dot hybridization. (5–7) Establishment of virus-free cultures on MS medium, root induction, and transplantation to pots. (8) Maintenance of virus-free plants in an insect-proof greenhouse. (9) Field propagation of virus-free mother plants

3.2 Preparation of Pear Plants

3.2.1 Preparation of Potted Young Plants

1. Use 1-year-old virus-free seedlings in 20 cm × 25 cm pots as rootstocks (*see* **Note 1**).
2. Graft dormant shoots with 2–3 buds onto rootstocks in spring, or graft 2–3 buds onto rootstock stems in August to September.
3. Remove grafting bands after scion survival; cut rootstock shoots 0.5–1.0 cm above grafted buds in early spring (before bud break).
4. Prune shoot tips when scions develop 4–5 leaves to promote 2–3 young shoots; use plants with 3–5 shoots for thermotherapy.

3.2.2 Preparation of In Vitro Cultured Plants

All operations are performed in a Clean Workbench except **steps 1** and **2**.

1. Excise ~1 cm shoot tips from candidate plants in spring.
2. Remove leaves from shoot tips using a sharp scalpel.
3. Disinfect shoots by sequential rinsing in 70% ethanol (5–10 min), sterile water (3 washes), 3% sodium hypochlorite (5–10 min), and sterile water.
4. Excise 5 mm tips and transfer to MS medium in 100 mL glass flasks containing 50 mL medium (4–5 shoots per flask).
5. Incubate cultures in a growth room at 24 ± 1 °C under a 16 h photoperiod with 2000 L × light intensity.
6. Subculture ~1 cm shoots onto fresh MS medium at 30-day intervals (3–4 times) to generate sufficient material for virus elimination treatments (*see* **Note 3**).

3.3 Thermotherapy of Potted Plants

1. Place potted candidate plants (2–3 shoots with 3–5 leaves each) into a thermotherapy chamber in a glass growth room (*see* **Note 2**).
2. Set temperature to 37 °C and maintain normal irrigation (*see* **Note 6**).
3. At 30–35 days posttreatment, excise 0.5–1 cm young shoots for grafting onto 1-year-old virus-free potted rootstock (*see* **Note 7**; Fig. 2), or culture 0.2–0.5 cm shoot tips on MS medium.
4. Screen treated plants for virus presence by ELISA, dot–blot hybridization, or RT-PCR (*see* **Notes 8** and **9**).

3.4 Thermotherapy of In Vitro Plants

Excise ~7 mm shoots from in vitro cultures and transfer to MS medium (prepare ≥25 plants in five flasks per cultivar to ensure adequate survival). Maintain cultures in a growth room for 2 days, then transfer to growth chambers. After a 4-day pre-treatment at 32 °C with a 16 h photoperiod, apply thermotherapy as follows:

Fig. 2 Micrografting process and resulting plants. (1) Heat-treated shoot tips undergoing micrografting, covered with plastic bags. (2) Surviving shoots post-grafting. (3) Virus-free plants derived from thermotherapy combined with micrografting

3.4.1 Constant 37 °C Treatment

1. Incubate cultures at 37 °C with a 16 h photoperiod for 30–35 days.
2. Dissect 1.0–5.0 mm meristem tips from treated shoots and transfer five tips per flask to fresh MS medium.
3. Culture at 24 °C with a 16-h photoperiod. Subculture newly developed shoots every 4–5 weeks (up to three times) under the same conditions. Label cultures as X-Y-Z (candidate plant ID-culture ID-thermotherapy-derived culture ID).

3.4.2 Alternating 34 °C/42 °C Treatment

1. Incubate cultures for 40–50 days under alternating temperatures: 38 °C (16 h light) and 34 °C (8 h dark).
2. Dissect and culture meristem tips as described in Subheading 3.2.2.
3. Subculture as described in Subheading 3.2.2.
4. Screen all cultures for virus presence by ELISA, dot-blot hybridization, or RT-PCR.

3.5 Transplanting of Virus-Free Plants

1. Excise 1–1.5 cm shoots from 30-day-old cultures.
2. Transfer three shoots per 100-mL flask containing 50-mL rooting medium (*see* **Note 10**).
3. Move rooted plantlets to a room or greenhouse at 20–30 °C under natural scattered sunlight for 1 week.
4. Open vessel caps and add sterile water to 0.5 cm above the medium surface.
5. Remove plantlets from vessels and rinse gently to wash off medium gel.
6. Transplant into moist nutrient soil in nursery beds or pots; cover with white plastic film to maintain >80% relative humidity.

7. Remove films and acclimatize plantlets at 20–30 °C under natural scattered sunlight.
8. Maintain established plants in an insect-proof greenhouse as virus-free germplasm for propagation.

4 Notes

1. *Pyrus betulifolia* and *P. calleryana* are common pear rootstocks in China. Seeds from virus-tested mother plants are sown in pots or nursery seedbeds and grown in a greenhouse.
2. The chamber is constructed with double-layer glass walls and ceiling for light transmission and temperature stability, with an axial flow fan for air circulation.
3. For sub-culturing, excise 0.5–1.0 cm shoots from established cultures and transfer to fresh MS medium under the same initial conditions.
4. Dissolve 3.73 g EDTA and 2.78 g FeSO4 separately in 450-mL distilled water with heating and stirring. Mix the solutions and adjust to 1 L in a brown glass bottle.
5. IBA and NAA concentrations vary by genotype. For example, optimal rooting medium for Pyrus pyrifolia cv. 'Jinshui No. 2' is 1/2 MS with 20 g/L sucrose, 5.8–6.0 g/L agar, 2.5 mg/L IBA, and 1.5 mg/L NAA; for P. communis cv. 'Conference', use 0.5 mg/L IBA and 1.5 mg/L NAA.
6. Irrigate regularly during treatment to maintain soil moisture and ensure normal plant growth.
7. Maintain high humidity with white plastic bags; remove after two weeks. Test surviving plants for viruses 2–3 times at 1-month intervals.
8. Smaller meristem tips increase virus elimination efficiency but reduce culture survival. Conduct virus detection after 2–3 subcultures.
9. *Apple chlorotic leaf spot virus* (ACLSV, *Trichovirus*), *Apple stem pitting virus* (ASPV, *Foveavirus*) and *Apple stem grooving virus* (ASGV, *Capillivirus*) commonly infect commercial pear cultivars. These viruses are detected by ELISA [18], dot-blot hybridization [19], or (multiple) RT-PCR [20]; see our previous reports for detailed protocols. For RNA extraction, see ref. [21].
10. Ensure 2–3 mm of shoot stem base remains in the medium for root induction. Record root number and length every 10 days during a 40-day induction period.

Acknowledgments

This work was supported by Chinese Ministry of Agriculture, with the Project nyhyzx-201203076-03, the earmarked fund for Pear Modern Agro-industry Technology Research System CARS-29-10.

References

1. Bettoni JC, Wang MR, Li JW, Fan XD, Fazio G, Hurtado-Gonzales OP, Volk GM, Wang QC (2024) Application of biotechniques for in vitro virus and viroid elimination in pome fruit crops. Phytopathology 114:930–954
2. Grout BWW (1999) Meristem-tip culture for propagation and virus elimination. In: Hall RD (ed) Plant cell culture protocols, methods in molecular biology, vol 111. Humana Press
3. Wang PJ, Hu CY (1980) Regeneration of virus-free plants through in vitro culture. In: Fiechter A (ed) Advances in biochemical engineering, vol 18. Springer, Berlin, Heidelberg
4. Bhat AI, Rao GP (2020) Virus elimination by meristem-tip culture. In: Characterization of plant viruses. Springer Protocols Handbooks. Humana, New York
5. Kassanis B (1957) Effects of changing temperature on plant virus diseases. Adv Virus Res 4: 221–241
6. Mink GI, Wample R, Howell WE (1998) Heat treatment of perennial plants to eliminate phytoplasmas, viruses, and viroids while maintaining plant survival. In: Hadidi A, Khetarpal RK, Koganezawa H (eds) Plant virus disease control. APS Press, St. Paul
7. Walkey DGA, Cooper VC (1975) Effect of temperature on virus eradication and growth of infected tissue cultures. Ann Appl Biol 80: 185–190
8. Wang QC, Cuellar WJ, RajamA ki ML, Hirata Y, Valkonen JPT (2008) Combined thermotherapy and cryotherapy for efficient virus eradication: relation of virus distribution, subcellular changes, cell survival and viral RNA degradation in shoot tips. Mol Plant Pathol 9: 237–250
9. Wang LP, Wang GP, Hong N, Tan RR, Deng XY, Zhang H (2006) Effect of thermotherapy on elimination of apple stem grooving virus and apple chlorotic leaf spot virus for in vitro-cultured pear shoot tips. Hort Sci 41:729–732
10. Cooper VC, Walkey DGA (1978) Thermal inactivation of cherry leaf roll virus in tissue cultures of Nicotiana rustica raised from seeds and meristem tips. Ann Appl Biol 88:273–278
11. Hu GJ, Hong N, Wang LP, Hu HJ, Wang GP (2012) Efficacy of virus elimination from in vitro cultured sand pear (*Pyrus pyrifolia*) by chemotherapy combined with thermotherapy. Crop Prot 37:20–25
12. Manganaris GA, Economou AS, Boubourakas IN, Katis NI (2003) Elimination of PPV and PNRSV through thermotherapy and meristem-tip culture in nectarine. Plant Cell Rep 22:195–200
13. Stein A, Spiegel S, Faingersh G, Levy S (1991) Responses of micropropagated peach cultivars to thermotherapy for the elimination of prunus necrotic ringspot virus. Ann Appl Biol 119: 265–271
14. Sharma S, Singh B, Rani G, Zaidi AA, Hallan VK, Nagpal AK, Virk GS (2008) In vitro production of Indian citrus ringspot virus (ICRSV) free kinnow plants employing phytotherapy coupled with shoot tip grafting. Plant Cell Tissue Organ Cult 92:85–92
15. Wang QC, Panis B, Engelmann F, Lambardi M, Valkonen JPT (2009) Cryotherapy of shoot tips: a technique for pathogen eradication to produce healthy planting materials and prepare healthy plant genetic resources for cryopreservation. Ann Appl Biol 154:351–363
16. Tan RR, Wang LP, Hong N, Wang GP (2010) Enhanced efficiency of virus eradication following thermotherapy of shoot-tip cultures of pear. Plant Cell Tissue Organ Cult 101:229–235
17. Murashige T, Skoog F (1962) A revised medium for rapid growth and bioassays with tobacco tissue cultures. Physiol Plant 15:473–497
18. Hong N, Wang GP, Yu JM, Dong YF, Zhang ZP (1997) The isolation, purification and serological detection of apple stem grooving virus. Sci Agric Sin 30:6–12
19. Wang LP, Wang GP, Tan RR, Hong N (2006) Detection of three latent viruses in pear by hybridization with biotinylated cDNA probes. Acta Phytopathol Sin 36:488–493

20. Yao BY, Lu YC, Li L, Wang YX, Li QY, Wang GP, Hong N (2018) A sensitive nested multiplex RT-PCR assay for the simultaneous detection of three common viruses infecting pear plants. J Virol Methods 263:105–110

21. Yang F, Wang GP, Xu WX, Hong N (2017) A rapid silica spin column-based method of RNA extraction from fruit trees for RT-PCR detection of viruses. J Virol Methods 247:61–67

Chapter 12

Combining Chemical Treatment with Shoot Tip Culture for Plant Virus Eradication in Fruit Trees

Guojun Hu and Yafeng Dong

Abstract

Antiviral chemical compounds effectively eradicate viruses and viroids across diverse plant genera. Virus elimination via chemotherapy is typically combined with shoot tip culture, with efficacy depending on antiviral type and concentration, treatment duration, target virus, and host species. Here, we detail a protocol using ribavirin to eliminate viruses from apple and grapevine, covering preparation of antiviral media, in vitro culture, chemical treatment, shoot tip culture, and iterative virus detection throughout the process. Methods for rooting virus-free plantlets and their transplantation to soil are also provided.

Key words In vitro culture, Chemical treatment, Shoot tip culture, Rooting culture and transplantation

1 Introduction

Chemical treatment combined with shoot tip culture is a common method for virus elimination in plants. During this process, plantlets, shoot tips, and/or meristems are cultured on antiviral-containing medium for a defined period, after which newly developed shoots or shoot tips are transferred to antiviral-free medium [1–4]. Most antiviral chemicals are base or nucleoside analogs that induce lethal mutations in viral genomes [5]. In plants, chemotherapeutic agents primarily target host enzymes and fall into three main classes: inosine monophosphate dehydrogenase (IMPDH) inhibitors, purine biosynthesis inhibitors, and neuraminidase inhibitors [6–9]. Ribavirin (1-β-d-ribofuranosy l-1,2,4-triazole-3-carboxamide; trade name: Virazol) is an IMPDH inhibitor and synthetic purine nucleoside analog structurally similar to guanosine. It inhibits IMPDH activity via its active metabolite, thiazole-4-carboxamide adenine dinucleotide [10], and has proven effective against many plant viruses and viroids [11–18]. However, the primary limitation of this approach is the phytotoxicity of antiviral

Guanwei Wu and Xiaofei Cheng (eds.), *Plant Antiviral Resistance: Methods and Protocols*, Methods in Molecular Biology, vol. 3039, https://doi.org/10.1007/978-1-0716-5300-5_12,

reagents. High ribavirin concentrations (> 50 μg/mL) can damage in vitro plant shoots, with toxicity levels varying by cultivar [3]. Therefore, optimizing dose and duration for each host–virus system is essential.

Viral or viroid diseases severely impair fruit tree growth and development worldwide, causing substantial yield losses [19, 20]. For example, prunus necrotic ringspot virus (PNRSV) causes chlorosis, mosaic patterns, ringspot, stunting, leaf distortion, and necrosis in peach, plum, apricot, and cherry [21]. Grapevine leafroll-associated viruses induce downward curling, reduce shoot vigor and fruit yield, and produce fruit with lower sugar and anthocyanin content and higher acidity [22, 23]. Citrus exocortis viroid (CEVd) causes stunting, bark cracking, leaf abnormalities, petiole cracks, low fruit set, and poor tree performance in citrus [24]. With no effective filed treatments available, using healthy propagation material is crucial for disease control. Here, we provide a detailed protocol on highly effective virus eradication using ribavirin chemotherapy combined with shoot tip culture, using apple and grapevine as cases.

2 Materials

2.1 In Vitro Culture

1. In vitro apple and grapevine plantlets are used as the plant materials (*see* **Note 1**).
2. The basic medium: Murashige and Skoog (MS) medium (*see* **Note 2**).
3. Proliferation medium for apple: MS, 1 mg/L 6-benzylaminopurine, 0.1 mg/L 3-indole butyric acid, 30 g/L sucrose, and 7–9 g/L agar (*see* **Note 3**).
4. Rooting medium for apple: 1/2 MS, 0.3 mg/L IBA, 15 g/L sucrose, and 5–6 g/L agar.
5. Grapevine: 1/2MS, 0.5 mg/L indole-3-acetic acid, 1.0 mg/L Kinin, 30 g/L sucrose, and 7 ~ 9 g/l L agar.

   The pH is adjusted to 5.8 with 1 M NaOH.

2.2 Ribavirin Treatment and Virus Detection

Prepare all solutions using ultrapure water (prepared by purifying deionized water).

1. 1000 μg/mL ribavirin: weight 50 mg ribavirin in 50 mL ultrapure water (prepared by purifying deionized water). Mix and filter-sterilize using a filter (0.22 μm). Store at 4 °C.
2. Grinding buffer: 4 M guanidine thiocyanate, 0.2 M sodium acetate, 25 mM EDTA, 1 M potassium acetate, 2.5% PVP MW 40,000, 1% β-mercaptoethanol.

3. Nucleic acid binding buffer: 300 μL 6 M NaI, 150 μL ethanol, 25 μL autoclaved silica powder (*see* **Note 4**).
4. Washing buffer: 10 mM Tris–HCl pH 7.5, 0.5 mM EDTA, 50 mM NaCl, 50% ethanol.
5. M-MLV RT 5 × buffer: 250 mM Tris–HCl (pH 8.3, 25 °C), 375 mM KCl, 15 mM $MgCl_2$, 50 mM DTT.

3 Methods

3.1 Flowchart of Chemical Treatment Combined with Shoot Tip Culture

The procedure is schematically illustrated in Fig. 1. Shoots from vigorous, uniformly growing plantlets were excised and cultured on ribavirin-containing medium. Following incubation, apical and axillary shoots were transferred to antiviral-free medium. Once shoots developed into new plantlets, survivors were screened for virus presence; virus-free plantlets were rooted and transferred to substrate-filled trays before field planting.

3.2 Preparation of MS Medium Containing Ribavirin

1. Prepare the MS or ½ MS medium. Aliquot it into culture bottles (240 mL), with 40 mL per bottle. Autoclave at 121 °C for 15–18 min.
2. Add the ribavirin stock solution to the medium when the medium cools to 60–70 °C (operate in clean workbench). Mix quickly and store the medium at room temperature.
3. The final concentration of ribavirin is 25 μg/mL. To achieve this, add 1 mL of 1000 μg/mL ribavirin stock solution to 40 mL MS medium (*see* **Note 5**).

Fig. 1 Technical flowchart (in vitro apple)

3.3 In Vitro Culture

1. Culture in vitro apple plantlets on full-strength MS medium and grapevine plantlets on half-strength MS medium.
2. Incubate all plant materials in tissue culture room at 24 °C, with a photoperiod of 16 h light and 8 h dark, and a light intensity of 2000 L× (*see* **Note 6**).
3. Transfer viable plantlets to fresh medium (operate on a clean workbench) at intervals of 30–40 days for apple and 60–70 days for grapevine. For apple, mainly excise the shoot tips, while for grapevine, excise both the shoot tips and segments.

3.4 Chemical Treatment (Operate in Clean Workbench)

1. Select strong and uniformly growing plantlets. Excise shoots from the in vitro plantlets, with lengths of approximately 1.0–1.2 cm for grapevine and 0.6–0.8 cm for apple.
2. Transfer the excised shoots to the ribavirin-containing medium. Place 4–6 shoots in each culture bottles.
3. Incubate all treated shoots in tissue culture room at 24 °C, with a photoperiod of 16 h light and 8 h dark, and a light intensity of 2000 L×. The incubation period should be 40–80 days.

3.5 Shoot Tip Culture (Operate in Clean Workbench)

1. At the end of the treatments, cut apical and axillary shoots that are approximately 1.0–2.0 mm in length, ensuring that the axillary shoots are at least 5 mm long.
2. Culture all shoot tips on appropriate medium (*see* **Note** 7). Incubate in tissue culture room at 24 °C, with a photoperiod of 16 h light and 8 h dark, and a light intensity of 2000 L×.
3. After 4–5 sub-culture cycles, test the regenerated plantlets for the presence of virus.

3.6 Virus Detection

3.6.1 RNA Extraction

1. Extract total RNA from plant samples following the method of Foissac et al. [25] with modifications: Homogenize 0.1 g of plant tissue in 1 mL of grinding buffer. Add 150 μL of 10% N-lauroyl sarkosyl to the homogenate (~0.6 mL).
2. Incubate the mixture at 72 °C for 10 min, then on ice for 5 min.
3. Centrifuge at 13,000 g for 10 min. Mix 300 μL of the supernatant with 300 μL of nucleic acid binding buffer. Wash the pellet twice with washing buffer.
4. Elute RNA in 100 μL of nuclease-free water. Incubate at 72 °C for 4 min, then centrifuge at 12,000 g for 3 min. Store the nucleic acids at −72 °C.

3.6.2 First-Strand cDNA Synthesis

1. Mix 1 μg of RNA with 5 μM random hexamer primers. Add nuclease-free water to a final volume of 10 μL.
2. Incubate at 72 °C for 10 min, then on ice for 5 min.

3. Add the following components to the mixture: 4 μL of M-MLV RT 5× buffer, 1 μL of 0.5 mM dNTP mix, 0.5 μL of 5 U/μL M-MLV reverse transcriptase, and 4.5 μL of nuclease-free water.
4. Incubate the mixture at 37 °C for 10 min, then at 42 °C for 1 h.
5. Terminate the reaction at 72 °C for 5 min, then on ice for 3 min.

3.6.3 PCR Amplification

1. Mix the following components: 2.5 μL of 10× PCR buffer with Mg^{2+}, 0.5 μL of 10 mM dNTP mix, 0.5 μL of each viral primer (final concentration 1 μM), 0.2 μL of 5 U/μL rTaq DNA polymerase, 2 μL of template cDNA, and18.8 μL of sterile water.
2. Pre-denaturation at 94 °C for 3 min, and run PCR in a 96-well PCR Thermal Cycler with 35 cycles of following steps: Denaturation at 94 °C for 30 s, annealing at 53 °C for 30 s, extension at 72 °C for 30 s, and final extension at 72 °C for 10 min.
3. Separate the amplicons using electrophoresis on 1.2% (w/v) agarose gels.
4. Stain the gels with a nucleic acid dye.
5. Visualize the results under ultraviolet light.

3.7 Rooting Culture

1. Proliferate virus-free in vitro plantlets of apple and grapevine on fresh medium.
2. Transfer the apple plantlets to the rooting medium (*see* **Note 8**).
3. Incubate all rooting plantlets in tissue culture room at 24 °C, with a photoperiod of 16 h light and 8 h dark, and a light intensity of 2000 Lx. The incubation period should be 40–80 days.

3.8 Transplantation of In Vitro Plantlets

3.8.1 Hardening

1. Transfer the rooted plantlets from the tissue culture room to a greenhouse or nursery.
2. Harden the plantlets 1–2 weeks (*see* **Note 9**). One day before transplanting, remove the vessel caps and add sterile water to soften the medium.

3.8.2 Transplantation

1. Gently remove the plantlets from their vessels.
2. Thoroughly rinse off any residual medium from the roots and stem bases (*see* **Note 10**).
3. Transfer the plantlets to nursery beds, pots, or trays filled with substrates (e.g., vermiculite, river sand, peat moss).

4. Use a tool to create planting holes or furrows, position the roots properly, and lightly compact the substrate around the stem base to stabilize the plantlets.
5. Water thoroughly using a fine mist to prevent dislodging. Replant any tilted seedlings immediately.
6. Maintain a relative humidity of >80% and a temperature 20–28 °C under plastic film and shade netting.
7. Gradually introduce ventilation and light exposure after approximately 10 days, once the seedlings have stabilized (*see* **Note 11**).

4 Notes

1. Suitable plant materials include plantlets, shoot tips, or meristems. Apple and grapevine cultivar were exemplified by *Malus* cv. Xinhong Jiangjun and *Vitis labrusca* × *V. vinifera* cv. Kyoho, respectively.
2. The basic medium can be MS, White, B5, N6, etc., depending on the type of plant or tissue being cultivated.
3. In vitro apples require different culture media for proliferation and rooting, while grapevines can use a single medium.
4. Ensure silica powder is mixed evenly before using the suspension solution.
5. Ribavirin concentrations above 50 mg/L in the culture medium may induce detrimental effects on in vitro plant shoots, with phytotoxicity varying by cultivar. Select an appropriate concentration as needed.
6. Adjust the temperature, light duration, and intensity according to the type of plants.
7. The medium should be fresh and free of ribavirin.
8. Root initiation of apple typically occurs within 7–10 days, followed by stem elongation within 10–15 days. After 1 month, plantlets develop into seedlings with 3–8 roots.
9. Exposure plantlets to natural light. Transplanting can being when leaves darken to a deep green and young stems turn purplish-red.
10. If roots are too long, they can be pruned appropriately.
11. Spring is the optimal season for transplantation.

Acknowledgments

This work was supported by the Agricultural Science and Technology Innovation Program of the Chinese Academy of Agricultural Sciences.

References

1. Bettoni JC, Wang MR, Li JW, Fan XD, Fazio G, Hurtado-Gonzales OP, Volk GM, Wang QC (2024) Application of biotechniques for *in vitro* virus and viroid elimination in pome fruit crops. Phytopathology 114:930–954
2. Karimpour S, Davarynejad G, Zakiaghl M, Safarnejad MR (2020) *In vitro* thermotherapy and thermo-chemotherapy approaches to eliminate some viruses in *Pyrus communis* L. cv. 'Natanz'. J Agric Sci Technol 22:1645–1653
3. Romadanova NV, Tolegen AB, Koken TE, Nurmanov MM, Kushnarenko SV (2021) Chemotherapy of apple shoots *in vitro* as method of virus eradication. Int J Biol Chem 14:48–55
4. Sedlák J, Pribylová J, Kolonuk I, Špak J, Lenz O, Semerák M (2023) Elimination of solanum nigrum ilarvirus 1 and apple hammerhead viroid from apple cultivars using antivirals ribavirin, rimantadine, and zidovudine. Viruses 15:1684
5. Díaz-Martínez L, Brichette-Mieg I, Pineño-Ramos A, Domínguez-Huerta G, Grande-Pérez A (2018) Lethal mutagenesis of an RNA plant virus via lethal defection. Sci Rep 8(1):1444
6. Monette PL (1983) Virus eradication through *in vitro* techniques. Int Plant Propag Soc 33: 90–100
7. Verma N, Ram R, Zaidi AA (2005) *In vitro* production of prunus necrotic ringspot virus-free begonias through chemo- and thermotherapy. Sci Hortic 103:239–247
8. Panattoni A, D'Anna F, Triolo E (2007) Antiviral activity of tiazofurin and mycophenolic acid against grapevine leafroll-associated virus 3 in *Vitis vinifera* explants. Antivir Res 73: 206–211
9. Luvisi A, Panattoni A, Triolo E (2011) Thiopurine prodrugs for plant chemotherapy purposes. J Phytopathol 159:390–392
10. Bougie I, Bisaillon M (2004) The broad spectrum antiviral nucleoside ribavirin as a substrate for a viral RNA capping enzyme. J Biol Chem 279:22124–22130
11. De Fazio G, Caner J, Vicente M (1978) Inhibitory effect of virazole (Ribavirin) on the replication of tomato white necrosis virus (VNBT). Arch Virol 58:153–156
12. Cieślińska M (2007) Application of thermo- and chemotherapy *in vitro* for eliminating some viruses infecting *Prunus* sp. fruit trees. J Fruit Orna Plant Res 15:117–124
13. El-Dougdoug KA, Osman ME, Abdelkader HS, Dawoud RA, Elbaz RM (2010) Elimination of hop stunt viroid (HSVd) from infected peach and pear plants using cold therapy and chemotherapy. Aust J Basic Appl Sci 4:54–60
14. Hu GJ, Dong YF, Zhang ZP, Fan XD, Ren F (2021) Elimination of grapevine fleck virus and grapevine rupestris stem pitting-associated virus from *Vitis vinifera* 87-1 by ribavirin combined with thermotherapy. J Integr Agric 20: 2463–2470
15. Hu GJ, Dong YF, Zhang ZP, Fan XD, Ren F, Li ZN, Zhang SN (2018) Elimination of grapevine rupestris stem pitting-associated virus from Vitis vinifera 'Kyoho' by an antiviral agent combined with shoot tip culture. Sci Hortic 229:99–106
16. Hu GJ, Hong N, Wang LP, Hu HJ, Wang GP (2012) Efficacy of virus elimination from in vitro cultured sand pear (*Pyrus pyrifolia*) by chemotherapy combined with thermotherapy. Crop Prot 37:20–25
17. Mahfouze SA, El-Dougdoug KA, Allam EK (2010) Production of potato spindle tuber viroid-free potato plant materials in vitro. J Am Sci 6:1570–1577
18. Savitri WD, Park KI, Jeon SM, Chung MY, Han JS, Kim CK (2013) Elimination of chrysanthemum stunt viroid (CSVd) from meristem tip culture combined with prolonged cold treatment. Hortic Environ Biotechnol 54:177–182
19. Walia Y, Dhir S, Bhadoria S, Hallan V, Zaidi AA (2012) Molecular characterization of apple scar skin viroid from Himalayan wild cherry. Forest Pathol 42:84–87
20. Sarropoulou V, Grigoriadou K, Maliogka VI, Sassalou CL, Ziogas V (2024) The elimination of viroids through *in vitro* thermotherapy and a meristem tip culture from a new limonime hybrid (*citrus* × *Limon* var. *Limon* (L.)

Burm. f. × *Citrus latifolia* var. *latifolia*). Biotech 13:37

21. Pallas V, Aparicio F, Herranz MC, Amari K, Sanchez-Pina MA, Myrta A, Sanchez-Navarro JA (2012) *Ilarviruses* of *Prunus* spp.: a continued concern for fruit trees. Phytopathology 102:1108–1120
22. Alabi OJ, Casassa LF, Gutha LR, Larsen RC, Henick-Kling T, Harbertson JF, Naidu RA (2016) Impacts of grapevine leafroll disease on fruit yield and grape and wine chemistry in a wine grape (*Vitis vinifera* L.) cultivar. PLoS One 11:e0149666
23. Halldorson MM, Keller M (2018) Grapevine leafroll disease alters leaf physiology but has little effect on plant cold hardiness. Planta 248:1201–1211
24. Bernard L, Duran VN, Elena SF (2009) Effect of citrus hosts on the generation, maintenance and evolutionary fate of genetic variability of citrus exocortis viroid. J Gen Virol 90:2040–2049
25. Foissac X, Svanella-Dumas L, Dulucq MJ, Candresse T, Gentit P (2001) Polyvalent detection of fruit tree tricho, capillo and foveaviruses by nested RT-PCR using degenerated and inosine containing primers (PDO RT-PCR). Acta Hortic 1:37–44

Chapter 13

Design Artificial Resistance Using Plant Elicitor Peptides

Xinyue Fan, Kaijing Li, Dangcheng Ji, Weiqin Ji, Xiaofei Cheng, and Xiaoyun Wu

Abstract

The lack of resistance resources seriously hampers disease-resistant breeding. The ability to customize artificial resistance has long been a dream for crop breeders. This chapter provides a detailed description of a recently reported an artificial resistance design using plant elicitor peptides. In theory, this artificial resistance strategy can easily be adapted to other protease-encoding pathogens by replacing the metacaspase cleavage site in plant elicitor peptides with the recognition sequence of the pathogen-encoded protease.

Key words Plant immunity, Molecular decoy, Protease

1 Introduction

Breeding disease-resistant crops is crucial for global food security. Current strategies primarily rely on either *resistance (R)* gene-mediated or *susceptibility (S)* gene-mediated immunity or resistance. However, the effectiveness of these strategies is limited by the scarcity of *R* and *S* genes, as well as the rapid resistance breaking by pathogens [1, 2]. In addition, both R and S proteins only confer specific resistance to typical pathogens and lack the flexibility to address emerging diseases or mixed infections. There is an urgent need to create modular, programmable artificial resistance strategies [3]. Recently, we reported an innovative strategy for customizing artificial resistance using plant elicitor peptides [4]. These peptides are usually expressed as precursors (ProPeps) in normal cells and are proteolytic cleaved by type-II metacaspases (MCs) into mature elicitor peptides (Peps) when the cell is damaged by injury or infection. The mature Peps are then secreted into the apoplast to activate plant innate immunity via cell surface-localized pattern recognition receptors (PRRs) [5, 6]. In the elicitor peptide-based artificial resistance, the MC cleavage site in ProPep is replaced with the recognition sequence of either the turnip mosaic virus

Guanwei Wu and Xiaofei Cheng (eds.), *Plant Antiviral Resistance: Methods and Protocols*, Methods in Molecular Biology, vol. 3039, https://doi.org/10.1007/978-1-0716-5300-5_13,

(TuMV)-encoded NIa-Pro or the *Pseudomonas syringae*-encoded AvrPphB [4]. TuMV or *P. suodomones* infection results in the cleavage of the engineered ProPep and activation of the resistance [4]. In theory, this artificial resistance strategy can be adapted to other protease-encoding pathogens. This "one-platform, multiple-resistance" design offers a remarkably simple yet flexible platform for customizing resistance to newly emerging pathogens. In this chapter, we provide a detailed experimental procedure for designing artificial resistance against other phytopathogens using TuMV as an example to promote sustainable crop production.

2 Materials

2.1 Plant Materials

1. Seeds of *Nicotiana benthamiana* and wild-type *Arabidopsis thaliana* ecotype Columbia-0 (Col-0) are maintained by the Molecular Plant Virology Laboratory at the Northeast Agricultural University.
2. The transgenic *A. thaliana* expressing *35S::proPep1*$^{NIa\text{-}Pro}$*-YFP* was described in our recent publication [4].

2.2 Virus, Bacteria, and Plasmids

1. *Escherichia coli* strain *DH5α*.
2. *Agrobacterium tumefaciens* strain *GV3101*.
3. The infectious clone of TuMV (pCambia-TuMV-6k2mCherry) was kindly provided by Dr. Aiming Wang from the London Research and Development Centre, Agriculture and Agri-Food Canada.
4. Plasmid pDONR207 was modified and preserved by the Molecular Plant Virology Laboratory.
5. Plasmid pEarley101 originates from the laboratory of Professor Keith W. Earley at Washington University in the United States.

2.3 Reagents and Materials

1. Antibiotics including Kanamycin (Kana), Gentamycin (Gent), Spectinomycin (Spec), and Rifampicin (Rif).
2. FastPure Plasmid Mini Kit.
3. FastPure Gel DNA Extraction Mini Kit.
4. LR Clonase™ II.
5. Eastep® Super Total RNA Extraction Kit.
6. HiScript III All-in-one RT SuperMix Perfect for qPCR.
7. HiScript III 1st Strand cDNA Synthesis Kit (+gDNA wiper).
8. ClonExpress II One Step Cloning Kit.
9. 2 × Phanta Max Master Mix.
10. 2 × Rapid Taq Master Mix.
11. ChamQ SYBR qPCR Master Mix.

12. One-step Plant Genome DNA Extraction Kit.
13. Agrobacterium Infiltration Buffer: 10 mM 2-Morpholinoethanesulphonic acid (MES) pH 5.6, 10 mM $MgCl_2$ and 10 μM acetosyringone (AS).
14. 0.08% Basta Solution: Dilute 1.6 mL of 20% Basta to a final volume of 400 mL.
15. Liquid LB medium: Weigh 10 g NaCl, 5 g yeast extract, 10 g typtone and dissolve in 1 L ddH_2O.
16. Silwet-77.

2.4 Laboratory Equipment

1. Stackable large-capacity constant temperature incubator shaker.
2. Plant tissue grinder.
3. Western blotting imaging system.
4. Plant growth chamber.
5. Eporator® electroporator.
6. pH meter.
7. Real-time PCR system.
8. Confocal microscope.
9. ImageJ software.

3 Methods

3.1 Construction of Engineered *proPep1* Vector

1. Retrieve the mRNA sequence of the *A. thaliana proPep1* gene (GenBank accession number: AT5G64900) from the NCBI database.
2. Purify total RNA from *A. thaliana* using the Eastep® Super Total RNA Extraction Kit.
3. Use the oligo-dT_{18} primer with the HiScript III 1st strand cDNA synthesis kit (+gDNA wiper) to reverse transcribe the total RNA into complementary DNA (cDNA).
4. Design the forward primer proPep1-F (5′- CAAAAAAG CAGGCTTCATGGAGAAATCAGATAG-3′) and the reverse primer proPep1-211R (5′-GGCCTGGTGATACACAAC-GACCTCCTCC-3′; the TuMV NIa-Pro cleavage site has been underlined) to amply 1–211 bp of the *proPep1* coding region from wild-type *A. thaliana* cDNA (*see* **Note 1**).
5. Design the forward primer proPep1-192F (5-′-CGTTGTGTATCACCAGGCCACAAAG-3′; the TuMV NIa-Pro cleavage site has been underlined) and the reverse primer proPep1-R (5′-GCTGGGTCCCGACGCTAATTA TGTTGGCC-3′) to amply the downstream (192–276 bp)

coding region of proPep1 from wild-type *A. thaliana* cDNA (*see* **Note 2**).

6. Use overlap extension PCR to join the two fragments obtained in **steps 4** and **5** (*see* **Note 3**).
7. Linearize the pDONR207 vector by PCR using the 2 × Phanta polymerase, the forward primer pDONR207-F (5′-CGTCGGG ACCCAGCTTTCTTG-3′) and the reverse primer pDONR207-R (5′-CATGAAGCCTGCTTTTTTGTAC-3′).
8. Analyze the PCR products from **steps 6** and **7** on a 1% agarose gel, excise the target bands, and purify them using a DNA gel extraction kit (*see* **Note 4**).
9. Assemble the purified fragments using the Clonexpress II one step cloning kit to generate pDONR207-proPep1$^{NIa\text{-}Pro}$ construct.
10. Transform all ligation products into *E. coli* (*see* **Note 5**).
11. Screen positive clones by colony PCR using proPep1-F and proPep1-R primers.
12. Extract plasmid using the FastPure Plasmid Mini Kit and confirming by Sanger sequencing (*see* **Note 6**).
13. Use the LR Clonase-mediated reaction to transform the engineered proPep1$^{NIa\text{-}Pro}$ fragment into the pEarley101 vector to generate the *35S::*proPep1$^{NIa\text{-}Pro}$-YFP construct.
14. Transform all ligation products into *E. coli* (*see* **Note 5**).
15. Screen the correct clones by colony PCR using the 35SF primer (5′-CAAAGCAAGTGGATTGATGTGACATC-3′) and the proPep1-R primer.
16. Confirm the plasmid by Sanger sequencing (*see* **Note 6**).

3.2 Generation of Transgenic Plants

1. Transform *35S::*proPep1$^{NIa\text{-}Pro}$-YFP into the *A. tumefaciens* strain GV3101 via electroporation (*see* **Notes 7** and **8**).
2. Screen the correct clones by colony PCR using the 35SF and proPep1-R primers.
3. Culture the bacteria in 500 mL of liquid LB medium containing 50 μg/mL Kana and 25 μg/mL Rif at 28 °C overnight.
4. Harvest the bacteria by centrifugation at 6000 rpm for 3 min at room temperature (*see* **Note 9**).
5. Discard the supernatant and resuspend the bacterial pellets in 500 mL of a freshly prepared 5% (w/v) sucrose solution containing 0.03% Silwet L-77 (Add silwet 77 at a ratio of 10,000:3).
6. Adjust the optical density at 600 nm ($O.D._{600}$) to 1.0 (*see* **Note 10**).
7. Prepare robust wild-type *A. thaliana* with abundant floral inflorescences (*see* **Note 11**).

8. Remove all siliques using medical scissors (*see* **Note 12**).
9. Add the *Agrobacterium* cell suspension to a beaker.
10. Submerge all inflorescences in the bacterial suspension for 1–2 min.
11. Lay the plants horizontally after submerge, secure the inflorescences with stakes, cover with cling film and keep in darkness for 18–24 h.
12. Remove the wrap and return the plants to the growth chamber to await seed maturation.
13. Collect the seeds when they are mature.
14. Sow the mature T0 generation seeds uniformly in soil mixed with sand at a ratio of 1:3 ratio.
15. Stratify the pots at 4 °C for 48 h, then transfer them to the growth chamber.
16. Spray the seedlings three times with 0.08% Basta (glufosinate ammonium) at one-day intervals (*see* **Note 13**).
17. Observe the plants for 5–8 days after the final spray.
18. Harvest the seeds from self-pollinated T1 plants individually and then screen the T2 seeds using 0.08% Basta again.
19. Select T2 plants from lines showing the 3:1 segregation ratio and harvest the seeds individually.
20. Select surviving transgenic plants (T3) showing vigorous growth for subsequent validation once homozygosity is obtained (*see* **Note 14**).

3.3 Evaluate the Resistance of Transgenic Plants to TuMV

1. Sow wild-type and homozygous transgenic plants in soil mixed with sand at a ratio of 1:3.
2. Agroinfiltrate two lower, fully expanded leaves of 4-week-old wild-type and *35S::proPep1*$^{NIa\text{-}Pro}$*-YFP* transgenic plants with agrobacteria harboring TuMV-6K2mCherry.
3. Maintain the inoculated plants in a controlled growth chamber and monitor them daily for symptom development from 7 days post-inoculation (dpi; *see* **Note 15**).
4. Capture bright-field images of the plants against a black background. Acquire fluorescence images using the appropriate wavelength for mCherry (*see* **Note 16**).
5. Analyze the images using ImageJ software to quantify the total leaf area and the fluorescent, virus-infected area. Then calculate the ratio of the infected area to the total leaf area.
6. Grind the samples using a mortar and pestle.
7. Extract RNA using the Eastep® super total RNA extraction kit, reverse transcribe into cDNA using the HiScript III all-in-one RT superMix.
8. Detect the expression levels of the TuMV *Coat protein* (*CP)* gene via quantitative PCR (qPCR) (*see* **Note 17**).

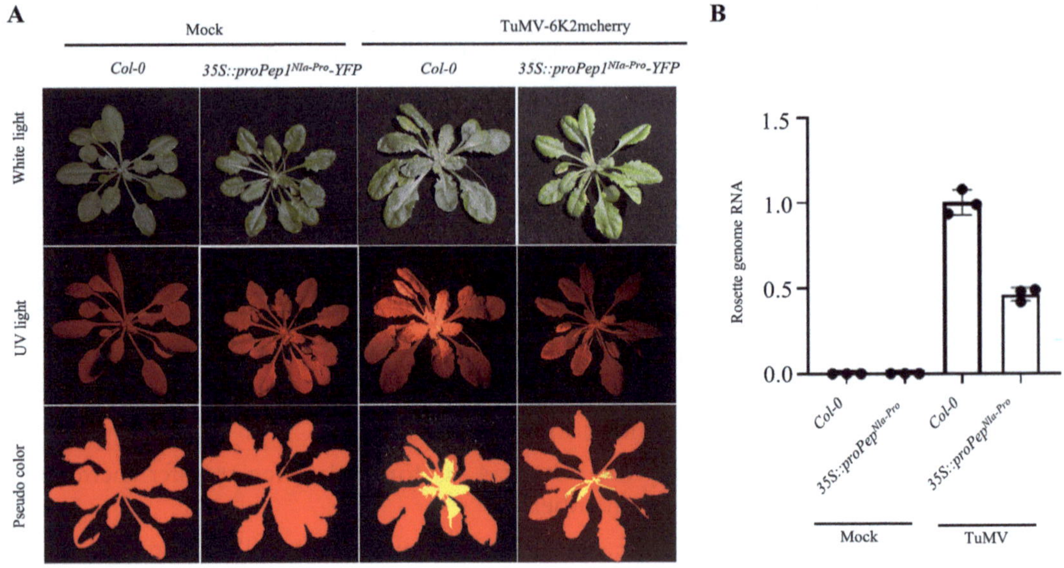

Fig. 1 Evaluate the resistance of engineered elictor peptide. (**a**) Phenotype of *Col-0* and *35S::proPep1*$^{Nla\text{-}Pro}$-*YFP* plants after TuMV infection under white and UV light at 11dpi. (**b**) RT-qPCR analysis of TuMV genomic RNA accumulation in the inoculated plants at 11 dpi

3.4 Expected Results

Wild-type *A. thaliana* seedlings display typical TuMV symptoms, including chlorosis and mosiac at 11 days post-inoculation (dpi), whereas transgenic plants exhibit only very mild symptoms of the virus (Fig. 1a). Moreover, RT-qPCR results reveal that transgenic *A. thaliana* seedlings accumulate significantly less of the viral genome than wild-type plants at 11 dpi (Fig. 1b). Taken together, these data suggest that transgenic plants are more resistance to TuMV infection.

4 Notes

1. For better results, increase the number of thermal cycles in the PCR to 35.
2. Due to the short size of the amplicons, the PCR product can be used directly for the subsequent overlap extension PCR.
3. The ratio of the two fragments should be calculated based on their respective concentrations and lengths.
4. When excising the target band, ensure that it is single and bright.
5. *E. coli DH5α* should not be shaken too violently, as this will affect the transformation efficiency.
6. The plasmid concentration must be greater than 100 ng/μL for normal sequencing.

7. Check the electrode cup for cracks before use.
8. Before placing the electrode cup into the electroporator, make sure it is free of water or ice.
9. Bacterial pellets should be gently resuspended to avoid cell death.
10. After adding AS, incubate the bacterial pellets at room temperature in the dark for at least 1 h.
11. Select robust and vigorous *A. thaliana* to obtain a greater quantity of seeds. It is also necessary to prune off the first inflorescence in order to produce more inflorescence.
12. Take care to avoid damaging the inflorescence.
13. Basta should be sprayed evenly onto each seedling to ensure effective screening.
14. Strains with high proPep1-YFP expression levels should be selected by Western blotting for subsequent experiment.
15. TuMV-6K2mCherry typically develops a systemic infection at 7 dpi under normal conditions, so the optimal imaging time should be selected.
16. Bright-field and ultraviolet-field images of the plants must be captured from the same position.
17. The *A. thaliana EF1α* gene (GenBank accession number: At1g07920) is used as the internal control.

Acknowledgments

This work is financial supported by the Natural Science Foundation of Heilongjiang Province (TD2022C003), the National Natural Science Foundation of China (32470152) and the Leading Talents Supporting Program of the Northeast Agricultural University (NEAU2023QNLJ-010).

References

1. Kourelis J, van der Hoorn RAL (2018) Defended to the nines: 25 years of resistance gene cloning identifies nine mechanisms for R protein function. Plant Cell 30:285–299
2. Jones JD, Staskawicz BJ, Dangl JL (2024) The plant immune system: from discovery to deployment. Cell 187:2095–2116
3. Wu M, Zhao S, Wu J (2025) Simple but effective: minimalist NLR rewiring, maximal virus resistance. Mol Plant 18:1803–1805
4. Fan X, Zhao Y, Ji W et al (2025) A pathogen protease-activated molecular decoy for customized resistance in plant. Plant Biotechnol J 23:2403–2405
5. Shen W, Liu J, Li JF (2019) Type-II metacaspases mediate the processing of plant elicitor peptides in Arabidopsis. Mol Plant 12:1524–1533
6. Huffaker A, Dafoe NJ, Schmelz EA (2011) ZmPep1, an ortholog of Arabidopsis elicitor peptide 1, regulates maize innate immunity and enhances disease resistance. Plant Physiol 155: 1325–1338

Chapter 14

Virus-Induced *NLR*-Silencing Library-Based Screening to Identify Antiviral *NLR* Genes

Jiajia Lin, Jianping Chen, Fei Yan, and Guanwei Wu

Abstract

Traditional genetic methods for identifying disease-resistant nucleotide binding and leucine-rich repeat-containing receptors (NLRs) proteins are often laborious and time-consuming. To streamline the screening process, we employed a pre-constructed combinatorial Virus-Induced Gene Silencing (VIGS) library. This library targets 301 *NbNLR* genes by integrating specific fragments from 4 to 6 distinct *NLRs* into a single VIGS vector, generating 55 VIGS combination (VIGS-com). This chapter presents an efficient method based on tobacco rattle virus (TRV)-mediated VIGS for rapidly screening *NLR* genes that confer resistance to plant virus infection in *Nicotiana benthamiana*, using turnip mosaic virus (TuMV) as a case. This approach enables rapid and systematic screening of key disease-resistant NLR receptors in antiviral plant immunity.

Key words *LR* genes, TuMV, VIGS, Disease resistance

1 Introduction

The plant immune system is crucial for perceiving both pathogenic and endogenous danger signals. Plants utilize a vast array of cell-surface and intracellular immune receptors to detect diverse immune signals associated with pathogen infection [1]. Among these, *NLR* serve as key executors of disease resistance. They specifically recognize pathogen effectors and confer targeted resistance against pathogens at the site of infection [2]. The NLR protein family is extensive, so conventional genetic approaches such as map-based cloning typically involve constructing segregating populations, performing multigenerational genetic mapping, and conducting laborious complementation tests [3]. This process is often time-consuming and resource-intensive.

VIGS enables rapid, high-throughput downregulation of target genes through transcript degradation and is widely used for functional gene analysis [4]. TRV is one of the most extensively

Guanwei Wu and Xiaofei Cheng (eds.), *Plant Antiviral Resistance: Methods and Protocols*, Methods in Molecular Biology, vol. 3039,
https://doi.org/10.1007/978-1-0716-5300-5_14,

applied VIGS vectors. It overcomes limitations related to host range and meristem exclusion, enabling more efficient systemic spread throughout the plant while causing milder viral symptoms. In TRV-mediated VIGS, a fragment of the target gene is inserted into the pTRV2 vector. Upon inoculation, the TRV virus moves systemically and triggers the plant's RNA interference (RNAi) machinery, leading to sequence-specific degradation of endogenous mRNAs homologous to the inserted fragment [5, 6].

Here, we describe a detailed protocol for screening *NLR* genes conferring resistance to plant virus infection with TuMV as an example, using a recently reported TRV-VIGS-based *NLR* library from *Nicotiana benthamiana* [7]. This library consists of 301 *NbNLR* genes with unique 120-bp fragments and integrates 4–6 of these fragments into single vectors to form a high-efficiency screening system of 55 VIGS combinations (VIGS-com).

2 Materials

2.1 Plasmids

1. The *N. benthamiana NLR* VIGS library comprises 55 VIGS-com, with each VIGS-com containing silencing fragments targeting 4–6 distinct *NLR* genes [7].
2. pCB301 harboring the TuMV infectious clone.
3. pYL192 (pTRV2), a TRV infectious clone vector for systemic silencing.
4. pTRV2-PDS plasmid containing a fragment of phytoene desaturase (PDS).
5. pTRV1 plasmid containing the TRV RNA1 segment.

2.2 Plant Materials and Bacterial Strains

1. 2- to 4-week-old *N. benthamiana* plants (*see* **Note 1**).
2. *Escherichia coli* DH5α strain.
3. *Agrobacterium tumefaciens* strain GV3101.

2.3 Equipment

1. Thermal cycler with gradient functionality.
2. Electrophoresis apparatus.
3. Incubator shaker.
4. Thermostatic incubator.
5. Electroporator.
6. Microvolume spectrophotometer.
7. Centrifuge.

2.4 Reagents

1. Kanamycin.
2. Rifampicin.
3. FastDigest PstI (with buffer).

4. dTTP and dATP.
5. T4 DNA polymerase (with buffer).
6. Gel Extraction Kit.
7. Plasmid Extraction Kit.

2.5 Mediums and Buffers

1. LB liquid medium: Dissolve 5 g yeast extract, 10 g tryptone, and 10 g NaCl in 800 mL distilled water. Adjust the pH to 7.0 with NaOH, bring the final volume to 1 L with distilled water, and autoclave at 121 °C for 20 min.
2. LB solid medium: Add 15 g/L agar to the LB liquid medium prior to autoclaving, sterilize by autoclaving.
3. Agrobacterium infiltration buffer: 10 mM Na_3PO_4, 10 mM MES, 200 μM As, 0.5 mg/mL glucose. Adjust the pH to 5.6 with HCl. The solution can be filter-sterilized or autoclaved (*see* **Notes 2** and **3**).
4. Phosphate-buffered saline (PBS, 0.1 M): Dissolve 8.0 g NaCl, 0.2 g KCl, 0.259 g KH_2PO_4, and 2.17 g $Na_2HPO_4{\cdot}7H_2O$ in 800 mL distilled water. Adjust the pH to 7.4 with HCl, then bring the final volume to 1 L with distilled water. Sterilize by autoclaving.
5. Plant virus inoculation buffer: 0.01 M PBS supplemented with 2% (w/v) Na_2SO_3.

3 Methods

3.1 Flowchart

A brief schematic diagram of the entire procedure is illustrated in Fig. 1. Briefly, plasmids containing the TRV-VIGS-com were transformed into *Agrobacterium tumefaciens* for infiltration of *Nicotiana benthamiana* leaves. At 10–12 days post-infiltration (dpi), plants were mechanically inoculated with a fluorescently tagged test virus to monitor systemic infection under UV illumination. Differences in fluorescence enabled identification of antiviral VIGS-com constructs (Fig. 1a). Individual TRV-VIGS vectors targeting single *NLR* genes were then generated and screened using the same procedure to identify the specific *NLR* gene conferring the antiviral phenotype (Fig. 1b).

3.2 NLRcom-VIGS Assay

1. Introduce the following plasmids separately into GV3101 via electroporation: the pTRV2-NLRcom from the VIGS library, pTRV1, empty pTRV2, and the positive control pTRV2-PDS. Plate the transformations on LB solid medium supplemented with kanamycin and rifampicin, and incubate the inverted plates at 28 °C for 48 h.
2. Confirm positive clones by colony PCR.

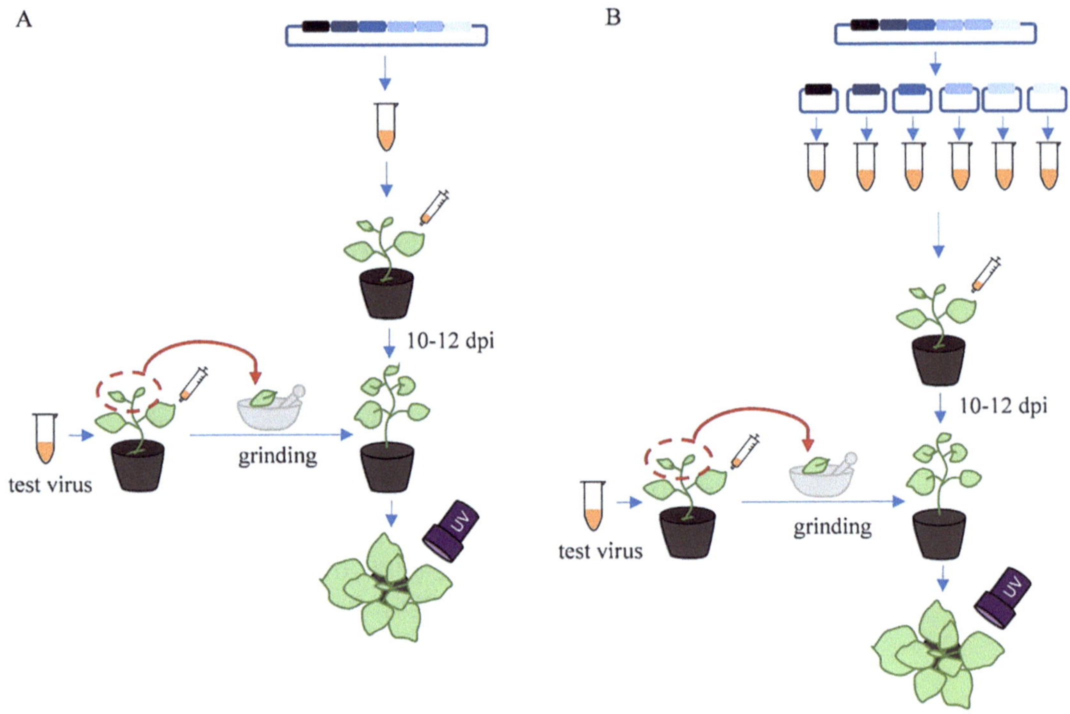

Fig. 1 Overview of VIGS-based antiviral *NLR* screening strategies. (**a**) High-throughput candidate screening using VIGS-com. (**b**) Individual *NLR* gene validation within the VIGS-com system

3. Inoculate a single positive colony into 700 μL of LB liquid medium containing kanamycin and rifampicin. Shake at 200 rpm, 28 °C overnight. Then, transfer the culture to 5 mL of the same antibiotic medium and incubate overnight under the same conditions.
4. Harvest the bacterial cells by centrifugation at 5000 rpm for 5 min. Wash the pellet once with 1 mL of sterile distilled water and then three times with 1 mL of Agrobacterium infiltration buffer, with centrifugation steps between each wash. Finally, resuspend the final pellet in infiltration buffer.
5. Mix the pTRV1 bacterial suspension individually with suspensions of empty pTRV2, pTRV2-NLRcom, or pTRV2-PDS and adjust the final OD_{600} of the Agrobacterium to 0.1 using Agrobacterium infiltration buffer.
6. Infiltrate the mixtures into the leaves of 2- to 3-week-old *N. benthamiana* plants using a 1 mL needleless syringe.

3.3 Secondary Virus Inoculation

Perform mechanical inoculation with tested virus at 10–12 days post-TRV infiltration.

1. Prepare an Agrobacterium culture containing the tested virus infectious clone as described in Subheading 3.2, and adjust the OD_{600} to 0.1.

2. Infiltrate this suspension into the leaves of 3- to 4-week-old *N. benthamiana* plants using a 1 mL needleless syringe.
3. Harvest virus-infected systemic leaves for virus inoculum source.
4. Grind the leaf tissue in a pre-chilled mortar with liquid nitrogen. Add 10 mL of plant virus inoculation buffer per 0.2 g leaf tissues and homogenize thoroughly. Clarify the extract by centrifugation at 12,000 rpm for 2 min at 4 °C. Keep the supernatant (viral inoculum) on ice (*see* **Note 4**).
5. Dust the third and fourth systemic leaves of TRV-pre-infiltrated plants with 200–300 mesh carborundum (*see* **Note 5**). Pipette 100 μL of the viral inoculum onto the dusted leaf surface and gently rub the leaf in one direction 6–7 times. After 2–3 min, gently rinse the leaves with distilled water to remove excess abrasive and inoculum.

3.4 Virus Infection Phenotype Evaluation

Upon systemic spread of the virus to the upper leaves of *N. benthamiana*, assess the fluorescence intensity and area in systemic leaves under UV light. Subsequently, employ molecular biology techniques (e.g., RT-qPCR) to validate the findings. VIGS-com constructs that, upon silencing, enhance virus infection are identified as candidates involved in virus resistance.

3.5 Construction of Individual TRV-VIGS Vectors Targeting Single NLR Genes

1. Digest the pYL192 vector using FastDigest PstI restriction enzyme, following the manufacturer's recommended reaction composition and incubation time.
2. Purify the linearized vector fragment using a gel extraction kit.
3. Assemble a reaction mixture containing 50 μL of the purified product, 10 μL of 10× buffer, 5 μL of 100 mM dTTP, 2 μL of T4 DNA polymerase, and 3.3 μL of distilled H_2O. Incubate at 37 °C for 20 min, followed by enzyme inactivation at 75 °C for 20 min.
4. Based on the library information, identify the 5–6 individual *NLR* genes from the antiviral candidate VIGS-com identified in the previous screen. Amplify the unique 120-bp VIGS fragment specific to each gene using a PCR thermal cycler.
5. Separate the PCR product via 1% agarose gel electrophoresis. Excise the gel slice containing the DNA fragment of the expected size.
6. Purify and recover the target PCR fragment from the gel using a gel extraction kit.
7. Prepare a tailing reaction mixture containing 5 μL of the purified insert, 1 μL of 10 × buffer, 0.5 μL of 100 mM dATP, 0.2 μL of T4 DNA polymerase, and 3.3 μL of ddH_2O. Incubate the mixture in a PCR instrument at 37 °C for 20 min, followed by 75 °C for 20 min.

8. Combine 3 μL of the dA-tailed product with 2 μL of the dT-tailed pTRV2 vector. Incubate at 37 °C for 20 min, then at 75 °C (with a ramp rate of 0.2 °C/s) for 20 min.
9. Add 5 μL of the ligation product to 50 μL of competent DH5α cells and incubate on ice for 30 min.
10. Transfer the cell suspension to a pre-heated water bath at 42 °C for 45 s.
11. Place the cell suspension on ice for 2 min, then add 700 μL of LB liquid medium.
12. Shake the cells at 200 rpm and 37 °C for 1 h. Plate the culture onto LB solid medium containing kanamycin for selection.
13. Identify positive single colonies by colony PCR.
14. Inoculate a positive colony into 5 mL of LB liquid medium supplemented with kanamycin. Grow an overnight liquid culture with shaking at 37 °C.
15. Isolate the plasmid using a plasmid miniprep kit and verify the sequence by Sanger DNA sequencing.
16. Transform the sequence-verified individual pTRV2-NLR plasmids into GV3101 following the procedure described in Subheading 3.2.
17. Repeat the Subheadings 3.3 and 3.4.

3.6 Expected Results

Here, we used TuMV as an example to exemplify this protocol. As shown in Fig. 2, control plants pre-infected with pTRV1 and pTRV2 showed systemic infection by TuMV-GFP at 6 dpi, while plants pre-infiltrated with certain pTRV2-NLRcom and pTRV1

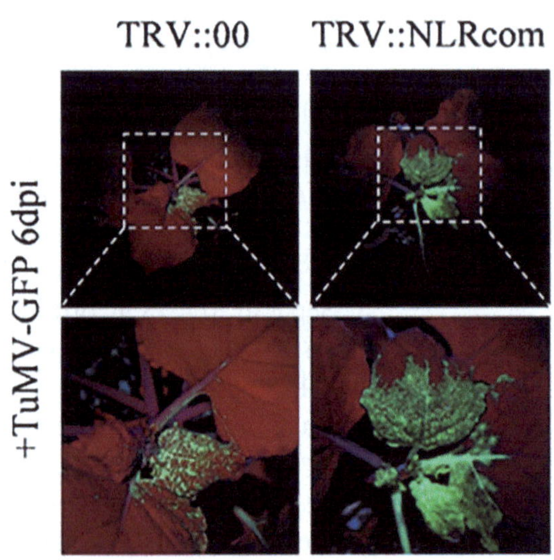

Fig. 2 Representative results of systemic leaf GFP fluorescence at 6 days post-inoculation with TuMV-GFP

exhibited a substantially stronger green fluorescence intensity, indicating that the NLRcom contains NLR gene(s) required for TuMV resistance. Subsequent VIGS targeting individual genes within this NLRcom of interest will identify the critical NLR gene (s) regulating TuMV resistance.

4 Notes

1. Water *N. benthamiana* plants thoroughly before inoculation to prevent drought stress from compromising plant health and confounding experimental results.
2. Adjust Agrobacterium infiltration buffer pH to 5.4–5.8. This mildly acidic range mimics wounded plant tissue and is essential for maximal bacterial virulence and activity.
3. Add heat-labile acetosyringone, a key *Agrobacterium virulence* inducer, to infiltration buffer after autoclaving and immediately before use to prevent oxidation and degradation.
4. Keep viral inoculum for mechanical inoculation on ice throughout preparation and until use to preserve viral particle integrity, prevent enzymatic degradation, and maintain high infectivity.
5. Sample for gene silencing efficiency assessment before viral challenge inoculation. This critical timing ensures baseline transcript measurements remain uncontaminated by virus-induced global transcriptional reprogramming, enabling accurate attribution of subsequent phenotypes to target NLR gene silencing.

Acknowledgments

The authors are indebted to Prof. Kee Hoon Sohn (Seoul National University) for kindly providing the *NbNLR* VIGS library plasmids. This research was funded by grants from the Zhejiang Provincial Natural Science Foundation (LZ24C140001) and National Natural Science Foundation of China (32370144) to G.W., Chinese Agriculture Research System of MOF and MARA (CARS-24-C-04) to F.Y., and the K. C. Wong Magna Fund in Ningbo University.

References

1. Zhou JM, Zhang Y (2020) Plant immunity: danger perception and signaling. Cell 181:978–989
2. Huang S, Jia A, Ma S et al (2023) NLR signaling in plants: from resistosomes to second messengers. Trends Biochem Sci 48:776–787

3. Dai Z, Pi Q, Liu Y et al (2024) ZmWAK02 encoding an RD-WAK protein confers maize resistance against gray leaf spot. New Phytol 241:1780–1793
4. Jagram N, Dasgupta I (2023) Principles and practice of virus induced gene silencing for functional genomics in plants. Virus Genes 59:173–187
5. Ratcliff F, Martin-Hernandez AM, Baulcombe DC (2001) Technical advance: tobacco rattle virus as a vector for analysis of gene function by silencing. Plant J 25:237–245
6. Burch Smith TM, Anderson JC, Martin GB et al (2004) Applications and advantages of virus-induced gene silencing for gene function studies in plants. Plant J 39:734–746
7. Ahn YJ, Kim H, Choi S et al (2023) Ptr1 and ZAR1 immune receptors confer overlapping and distinct bacterial pathogen effector specificities. New Phytol 239:1935–1953

Chapter 15

VIGS-Based Screening for *NLR* Genes Regulating Viral Effector-Induced Hypersensitive Response

Wenqi Mao, Jianping Chen, Fei Yan, and Guanwei Wu

Abstract

Nucleotide-binding leucine-rich repeat protein (NLR) can specifically recognize effector proteins secreted by pathogens, thereby initiating a strong immune response in host plants. Currently, utilizing this resistance gene to improve crop varieties is the most economically effective strategy. However, traditional identification methods are time-consuming and labor-intensive. RNA interference (RNAi) is a sequence-specific gene-silencing mechanism mediated by double-stranded RNA. This chapter provides a detailed protocol using tobacco rattle virus-mediated virus-induced gene-silencing (VIGS) screening platform in *Nicotiana benthamiana* for the fast identification of host *NLR* gene involved in viral effector-activated hypersensitive response (HR), with the turnip mosaic virus-coded nuclear inclusion protein a protease (NIa-Pro) as an example, which has been shown to activate HR in *Nicotiana benthamiana*.

Key words Nucleotide-binding leucine-rich repeat protein, Virus-induced gene silencing, Tobacco rattle virus, Hypersensitive response

1 Introduction

In the plant immune system, the nucleotide-binding leucine-rich repeat (NLR) protein family constitutes a key class of intracellular immune receptors responsible for specifically recognizing effector proteins secreted by pathogens, thereby activating effector-triggered immunity (ETI) [1, 2]. This immune response typically manifests as localized hypersensitive response (HR) and the initiation of systemic immune signaling [3], effectively suppressing pathogen infection and spread. The structure of NLR proteins generally consists of a central nucleotide-binding domain (NBD), a C-terminal leucine-rich repeat (LRR) domain, and an N-terminal signaling domain, enabling direct or indirect recognition of pathogen effectors. Typically, approaches to functionally test *Avr* gene candidates involve pairwise transient co-expression of individual candidate R and *Avr* gene combinations through polyethylene

Guanwei Wu and Xiaofei Cheng (eds.), *Plant Antiviral Resistance: Methods and Protocols*, Methods in Molecular Biology, vol. 3039, https://doi.org/10.1007/978-1-0716-5300-5_15,

glycol (PEG)-mediated protoplast transformation or leaf agroinfiltration to detect induced cell death [4, 5]. However, these methods assay candidate effectors one by one, requiring substantial labor and time.

RNA interference (RNAi) serves as a conserved gene silencing mechanism, initiated by Dicer-like (DCL) enzymes processing double-stranded RNA into small RNAs. These are incorporated into ARGONAUTE proteins within the RNA-induced silencing complex (RISC), directing either transcript cleavage or epigenetic repression [6]. Originally identified as an antiviral defense, RNAi now constitutes a fundamental genetic tool [7]. Virus-induced gene silencing (VIGS) exploits engineered plant viruses to trigger this RNAi pathway. Vectors are modified to remove pathogenicity while retaining replication and mobility functions, and are typically delivered via Agrobacterium or direct inoculation. They produce viral double-stranded RNAs in host cells, activating the host's silencing machinery to target homologous endogenous genes for knockdown, thereby providing a versatile platform for functional genomics [8]. Among VIGS vectors, those derived from Tobacco rattle virus (TRV) are the most extensively utilized. TRV, a typical member of the *Tobravirus* genus, possesses a bipartite single-stranded RNA genome (RNA1 and RNA2) separately packaged in distinct rod-shaped particles and replicates via a double-stranded RNA intermediate. TRV-based systems achieve effective gene silencing across Solanaceae species like *N. benthamiana*, tomato, pepper, potato, and petunia [9, 10]. A key advantage is TRV's ability to infect meristematic tissues, enabling research on flowering and fruit development [11]. TRV-VIGS methodology is also operationally straightforward, often employing Agrobacterium-mediated delivery via needle-free infiltration, soil drenching, or vacuum infiltration [12]. Recently, Ahn et al. constructed an *NbNLR* VIGS library comprising 55 silencing vectors targeting 301 *NbNLR* genes by assembling 4–6 VIGS fragments into each silencing vector, thereby reducing the number of vectors required for screening. This library serves as a valuable resource for identifying hypersensitive response (HR) associated with plant viral effector proteins.

TuMV, an important member of the genus *Potyvirus*, relies on the essential protease NIa-Pro for successful infection [13]. NIa-Pro can activate HR in two solanaceae plants, *Nicotiana benthamiana* and *Capsicum annuum* L [14]. In this chapter, we provided a detail protocol for screening and identification of NLR gene(s) involved in HR activation by viral effector protein, using TuMV NIa-Pro as a case.

2 Materials

2.1 Plasmids

1. The plasmids of pYL192 (a TRV infectious clone vector for systemic silencing).
2. The constructed silencing vector, pTRV1 (RNA1), pTRV2 (RNA2), pTRV2-PDS (phytoene desaturase).
3. The NbNLR VIGS library consists of 55 plasmid vectors, each designed from silencing fragment combinations to collectively target 301 NLR genes, was kindly provided by Prof. Kee Hoon Sohn from Seoul National University, Korea.

2.2 Plant Material and Bacterial Strains

1. 10- to 15-day-old *N. benthamiana* plants are grown in greenhouse at 22 °C with a light cycle of 16 h, a dark cycle of 8 h, and a relative humidity of 60% (*see* **Notes 1** and **2**).
2. Competent cells of *Escherichia coli strains* DH5α.
3. Competent cells of *Agrobacterium tumefaciens* strain GV3101.

2.3 Equipment and Software

1. Thermostatic incubator.
2. Autoclave machine.
3. 50 mL Falcon tubes.
4. pH meter.
5. 0.45 μm syringe filters.
6. Petri dishes.
7. Pipettes and tips.
8. 1.5 and 2.0 mL Eppendorf tubes.
9. Centrifuge 5420.
10. Vortex Oscillator.
11. Gene Pulser/MicroPulser Electroporation Cuvettes, 0.2 cm gap.
12. BioPhotometer.
13. Electrophoresis equipment.
14. High-speed refrigerated benchtop centrifuge.
15. Temperature gradient PCR instrument.
16. Handheld UV lamp.

2.4 Reagents and Buffers

1. Fast Digest PstI (with buffer).
2. 2 × Es Taq Master Mix.
3. Nucleic Acid Dye.
4. T4 DNA polymerase (with buffer).
5. PrimeSTAR Max DNA polymerase (with buffer).

6. Kanamycin.
7. Rifampicin.
8. Gel extraction kit.
9. Plasmid extraction kit.
10. LB liquid medium: Weigh 5 g yeast extract, 10 g peptone, 10 g NaCl into 800 mL distilled water and adjust the pH to 7.0 with NaOH, then volume to 1 L with distilled water and sterilize at 121 °C for 20 min in an autoclave.
11. LB solid medium: 1 L LB liquid medium is mixed with 15 g agar and autoclaved.
12. 1 M MES (2-(N-morpholino) ethanesulfonic acid)): dissolve 19.524 g MES in 100 mL distilled water, adjust the pH to 5.6 with NaOH, filter and sterilize, store at 4 °C.
13. 1 M $MgCl_2$ solution: dissolve 20.33 g $MgCl_2 \cdot 6H_2O$ in 100 mL distilled water. Filter sterilizes and store at 4 °C for spare parts.
14. 200 mM Acetosyringone (AS) solution: dissolve 39.24 mg AS in 1 mL dimethyl sulfoxide (DMSO). Store at −20 °C for spare parts.
15. Agrobacterium infiltration buffer (500 mL system): dissolve 5 mL of 1 M $MgCl_2$, 5 mL of 1 M MES, 500 μL of 200 mM AS in 500 mL of sterile water, mix well and store at 4 °C.

3 Methods

3.1 Preparation of NbNLR *VIGS Library*

1. Transfect the NbNLR VIGS library plasmids into *A. tumefaciens* GV3101 by electroshocking method.
2. Spread the bacterial suspension onto LB agar plates containing kanamycin and rifampicin, and incubate inverted at 28 °C in a constant temperature incubator for 48 h.
3. Verify the positive colonies by PCR.
4. Inoculate the positive strain into 5 mL LB liquid medium containing kanamycin and rifampicin and incubate overnight at 200 rpm in a constant temperature incubator at 28 °C.

3.2 Agroinfiltration of NLR *VIGS Library*

1. Centrifuge the cells at 5000 rpm for 5 min at room temperature to collect Agrobacterium cells.
2. Decant the supernatant and resuspend the precipitated cells with 1 mL agrobacterium infiltration buffer. Repeat the above two **steps 3–4** times to thoroughly remove residual antibiotics.
3. Mix pTRV1 separately with each bacterial suspension [pTRV2, pTRV-PDS, and pTRV2-NLRcom] and adjust the final

OD600 of each mixture to 0.1 using Agrobacterium infiltration buffer.

4. Agroinfiltrate these mixed bacterial suspensions into the leaves of 2- to 3-week-old *N. benthamiana* plants using a 1-mL needleless syringe.
5. Cultivate the plants in a greenhouse at 22 °C with 60% relative humidity under a 16-h light/8-h dark photoperiod for 10–12 days (*see* **Note 3**). pTRV2-PDS served as a positive control to evaluate silencing efficiency, as *PDS* silencing induces a visible photobleaching phenotype.

3.3 Viral Effector Expression and HR Phenotype Evaluation

1. Prepare *Agrobacterium* suspensions carrying the viral effector gene and *GUS* (β-glucuronidase) control according to the procedure in Subheading 3.2.
2. Adjust the OD_{600} to 0.1, and infiltrate the bacterial suspension into the different patches of the same silenced or control treatment leaves using a needleless 1-mL sterile syringe.
3. Cultivate plants in a greenhouse at 22 °C with 60% relative humidity under a 16-h light/8-h dark photoperiod for 3–4 days.
4. Monitor the HR phenotype development under both bright field and ultraviolet (UV) light conditions.
5. Select VIGS-NLRcom treatments that result in no or a weakened HR compared to the pTRV2 control (*see* **Notes 4** and **5**).

3.4 Construction of Individual NLR Gene-Silencing Vector and HR Phenotype Evaluation

1. Clone the silencing fragments for individual *NLR* genes identified from the candidate VIGS-NLRcom.
2. Insert each gene fragment to pTRV2 vector using a ligation independent cloning (LIC) method.
3. Repeat Subheadings 3.1, 3.2, and 3.3.
4. Critical *NLR* gene(s) required for viral effector-triggered hypersensitive response are identified based on phenotypic and quantitative differences from control plants.

3.5 Expected Results

Turnip mosaic virus-encoded NIa-Pro was used here to exemplify this protocol. NIa-Pro triggers HR in *N. benthamiana* [14], and its expression in pTRV1 + pTRV2 control plants induces robust HR (Fig. 1). In contrast, silencing specific *NLRcom* yields a substantially weakened HR phenotype under bright field and UV illumination (Fig. 1), indicating that the NLRcom contains *NLR* gene (s) required for NIa-Pro-mediated HR. Subsequent VIGS targeting individual genes within this NLRcom of interest will identify the critical NLR gene(s) regulating this response.

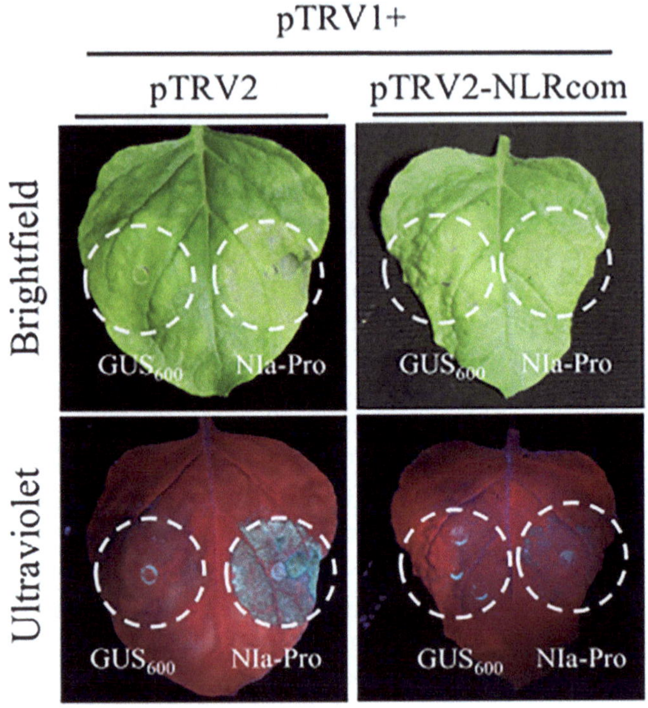

Fig. 1 Representative results of HR phenotypes triggered by TuMV NIa-Pro in pTRV2-NLRcom-silenced versus control *N. benthamiana* leaves

4 Notes

1. Healthy and uniform *Nicotiana benthamiana* seedlings should be used. Inoculation is ideally performed when the cotyledons are fully expanded and the first true leaf is just emerging. Silencing efficiency decreases significantly in older plants, while younger seedlings are more susceptible to infiltration damage.
2. Maintain an optimal watering regime to prevent the plants from experiencing drought or water stress prior to inoculation.
3. The duration of silencing is not fixed and is adjusted based on the bleaching phenotype observed in the PDS-positive control plants.
4. The silencing effect mediated by VIGS is heterogeneous among individual plants and even across different tissues of the same plant. Therefore, it is essential to observe a sufficient number of plants and record both the penetrance and intensity of the phenotype.
5. Simultaneous silencing using NLR-com fragments may lead to inconsistent silencing efficiency, and in some cases, individual genes may remain unsilenced, suggesting a limitation of this approach.

Acknowledgments

The authors are indebted to Prof. Kee Hoon Sohn (Seoul National University) for kindly providing the NbNLR VIGS library plasmids. This research was funded by grants from the Zhejiang Provincial Natural Science Foundation (LZ24C140001) and the National Natural Science Foundation of China (32370144) to G. W., Chinese Agriculture Research System of MOF and MARA (CARS-24-C-04) to F.Y., and the K. C. Wong Magna Fund in Ningbo University.

References

1. Huang S, Jia A, Ma S et al (2023) NLR signaling in plants: from resistosomes to second messengers. Trends Biochem Sci 48:776–787
2. Mäkinen K, Aspelin W, Pollari M, Wang L (2023) How do they do it? The infection biology of potyviruses. Adv Virus Res 117:1–79
3. Contreras MP, Lüdke D, Pai H et al (2023) NLR receptors in plant immunity: making sense of the alphabet soup. EMBO Rep 24: e57495
4. Kanja C, Hammond-Kosack KE (2020) Proteinaceous effector discovery and characterization in filamentous plant pathogens. Mol Plant Pathol 21:1353–1376
5. Arndell T, Chen J, Sperschneider J et al (2024) Pooled effector library screening in protoplasts rapidly identifies novel *Avr* genes. Nat Plants 10:572–580
6. Bologna NG, Voinnet O (2014) The diversity, biogenesis, and activities of endogenous silencing small RNAs in Arabidopsis. Ann Rev Plant Biol 65:473–503
7. Eamens A, Wang MB, Smith NA, Waterhouse PM (2008) RNA silencing in plants: yesterday, today, and tomorrow. Plant Physiol 147:456–468
8. Rössner C, Lotz D, Becker A (2022) VIGS goes viral: how VIGS transforms our understanding of plant science. Annu Rev Plant Biol 73:703–728
9. Spitzer-Rimon B, Marhevka E, Barkai O et al (2010) *EOBII*, a gene encoding a flower-specific regulator of phenylpropanoid volatiles' biosynthesis in petunia. Plant Cell 22:1961–1976
10. Hartl M, Merker H, Schmidt DD, Baldwin IT (2008) Optimized virus-induced gene silencing in *Solanum nigrum* reveals the defensive function of leucine aminopeptidase against herbivores and the shortcomings of empty vector controls. New Phytol 179:356–365
11. Spitzer B, Zvi MMB, Ovadis M et al (2007) Reverse genetics of floral scent: application of tobacco rattle virus-based gene silencing in petunia. Plant Physiol 145:1241–1250
12. Burch-Smith TM, Schiff M, Liu Y, Dinesh-Kumar SP (2006) Efficient virus-induced gene silencing in Arabidopsis. Plant Physiol 142:21–27
13. Wu G, Fang X, Yu T et al (2024) Turnip mosaic virus pathogenesis and host resistance mechanisms in *Brassica*. Hortic Plant J 10:947–960
14. Jia Z, Rui P, Fang X et al (2025) Proteolysis of host DEAD-box RNA helicase by potyviral proteases activates plant immunity. New Phytol 245:1655–1672

Chapter 16

High-Throughput Screening of Antiviral Agents Using the Fluorescence-Based Local Lesion Inspection Assay

Liansheng Yu, Hongru Wang, Jia Liu, Weiqin Ji, Xiaoyun Wu, and Xiaofei Cheng

Abstract

A high-throughput antiviral agent screening system called the fluorescence-based local lesion inspection assay (FLLIA) has been developed. In this assay, leaves were mechanically inoculated with an engineered virus expressing a fluorescent protein. One half of each leaf was treated with a test chemical and the other half was treated with a buffer solution. The number and size of viral infection lesions were then imaged and quantified. A reduction in the number of viral infection lesions and/or decreased fluorescence intensity in the chemically treated group suggests a replication inhibition activity, while a reduction in the size of fluorescent lesions indicates a movement inhibition activity. This simple method can efficiently identify antiviral agents and primarily differentiate their action mechanisms. A step-by-step protocol of FLLIA was provided in this chapter.

Key words Antiviral agent screening, FLLIA, Tobacco mosaic virus, Image quantification, Local lesion

1 Introduction

Plant viruses represent one of the most serious threats to the production of food, economic, and horticultural crops worldwide [1–3]. Plant viral diseases cause crop yield losses of 10–30% on average, while the outbreaks of certain viral diseases can lead to total crop failure [4–7]. Currently, the management of plant viral diseases predominantly relies on breeding resistant cultivars, crop rotation, removing infected plants, repelling insects with insect-proof nets and traps, and applying pesticides [8]. However, these strategies have significant limitations in terms of practical application. These include a lack of antiviral genetic resources, a high mutation rate of viruses, long operational cycles, high costs, insecticide resistance, and environment pollution [9–11]. Thus, highly

Guanwei Wu and Xiaofei Cheng (eds.), *Plant Antiviral Resistance: Methods and Protocols*, Methods in Molecular Biology, vol. 3039, https://doi.org/10.1007/978-1-0716-5300-5_16,

effective, low-toxicity, and highly specific antiviral compounds are necessary to control plant viral diseases to ensure the stability and sustainability of agricultural production [12].

Due to its operational simplicity and low cost, the half-leaf local-lesion assay is commonly used for screening antiviral agents. This method evaluates compound activity by comparing the number of local lesions caused by a plant virus on its lesion host in the treated and control half-leaves following inoculation with the virus [10–12]. However, this method relies only on visual observation and cannot distinguish whether a compound inhibits viral replication or blocks viral intercellular movement. To overcome the technical limitation of the half-leaf local-lesion assay, this study introduces a new screening assay that uses a genetically engineered plant virus which stably expresses a fluorescent protein. Following rub-inoculation, the fluorescent local lesions caused by the virus are recorded. Image analysis software is then used to quantify the number, size, and intensity of the fluorescent lesions caused by viral infection. A reduction in the number of fluorescent lesions and decreased fluorescence intensity in the treated group indicates replication inhibition activity, while a reduction in fluorescent lesions size indicates movement inhibition activity. This system provides an accurate yet simple and powerful tool for screening antiviral agents. The aim of this chapter is to provide a step-by-step protocol for screening antiviral compounds using this system.

2 Materials

2.1 Biological Materials

1. *Nicotiana benthamiana* seeds (Laboratory of Molecular Plant Virology at Northeast Agricultural University).
2. Yellow fluorescent protein (YFP)-tagged infectious clone of tobacco mosaic virus (TMV-YFP) (Dr. Hada Wuriyanghan from Innar Mongolia Agricultural University) (*see* **Note 1**).
3. *Agrobacterium tumefaciens* (synonym *Rhizobium radiobacter*) strain GV3101.

2.2 Chemicals

1. Kanamycin.
2. Rifampicin.
3. 2-(N-morpholino) ethanesulfonic acid.
4. Acetosyringone.
5. Silicon carbide.

2.3 Equipment

1. Mortar and pestle.
2. Medical scissors.
3. Constant temperature incubation shaker.

4. Plant growth chamber.
5. Digital dry bath block.
6. Laboratory pH meter.
7. Electronic balances.
8. Medical ultra-low temperature freezer.
9. Medical refrigerator.
10. Laboratory refrigerator.
11. Ultra-low temperature freezer.
12. Vertical high-pressure steam sterilizer.
13. In vivo plant imaging system.

2.4 Medium and Buffers

1. 100 mg/mL Kana: Dissolve 1 g of kana powder in 10 mL of sterile ddH_2O. Sterilize the solution by filtration through a 0.22 μm Micron Syringe filter. Aliquot into sterile 1.5 mL microcentrifuge tubes and store at −20 °C.
2. 25 mg/mL Rif: Dissolve 250 mg of rif powder in 10 mL of DMSO. Sterilize the solution by filtration through a 0.22 μm Micron Syringe filter. Aliquot into sterile 1.5 mL microcentrifuge tubes and store at −20 °C, protected from light.
3. LB liquid medium: Dissolve 10 g of peptone, 5 g of yeast extract, and 10 g of NaCl in distilled water to a final volume of 1 L. Adjust the pH to 7.0. Autoclave at 121 °C for 15 min.
4. Rub-inoculation buffer:

   Solution A (1 M Na_2HPO_4): Weigh 17.907 g of anhydrous Na_2HPO_4 (or the appropriate mass of the hydrate form of Na_2HPO_4) and dissolve completely in ddH_2O in a 50 mL beaker. Adjust the final volume to 50 mL with ddH_2O.

   Solution B (1 M NaH_2PO_4): Weigh 7.801 g of anhydrous NaH_2PO_4 (or the appropriate mass of the hydrate form of NaH_2PO_4) and dissolve completely in ddH_2O. Adjust the final volume to 50 mL with ddH_2O.

   To prepare the rub-inoculation buffer, pipette 577 μL of Solution A and 423 μL of Solution B into a clean 15 mL centrifuge tube. Add 9 mL of ddH_2O to the tube. Mix thoroughly by vortexing. The resulting mixture is the required rub-inoculation buffer (approximately 0.1 M phosphate buffer, pH ~7.0).
5. 1 M MES Buffer: Weigh 9.76 g of MES powder and dissolve in approximately 40 mL of ddH_2O. Stir at room temperature until completely dissolved. Adjust the pH to 5.6 by dropwise addition of NaOH solution. Bring the final volume to 50 mL with ddH_2O. Sterilize by filtration through a 0.22 μm Micron Syringe filter. Aliquot into sterile tubes and store at 4 °C, protected from light.

6. 1 M $MgCl_2$ Solution: Weigh 10.165 g of $MgCl_2 \cdot 6H_2O$ powder and dissolve directly in 30 mL of ddH_2O. Stir until the solution is completely clear. Adjust the final volume to 50 mL with ddH_2O. Sterilize by filtration through a 0.22-μm Micron syringe filter and store at 4 °C.
7. 100 mM AS: Weigh 0.1962 g of AS and dissolve in DMSO to a final volume of 10 mL. Immediately aliquot the stock solution into sterile 1.5 mL microcentrifuge tubes using a precision pipette (e.g., 10 μL per tube). Store all aliquots at −20 °C, strictly protected from light.
8. Agroinfiltration buffer: Piper 500 μL of 1 M $MgCl_2$ solution and 500 μL of 1 M MES buffer (pH 5.6) into a suitable container. Add ddH_2O to approximately 45 mL, mix well, and then adjust the final volume to 50 mL with ddH_2O. Prepare fresh before use or store short-term at 4 °C.

2.5 Software

1. Lumazone LF software.
2. Lumazone LF DBoxFL Imaging Tools.
3. ImageJ (ij154-win-java8).

3 Methods

3.1 Flowchart of FLLIA

A brief schematic diagram of the entire procedure is illustrated in Fig. 1. Briefly, leaves of 4- to 5-week-old seedlings were mechanically inoculated by genetically engineered virus expressing fluorescent protein. The inoculated leaf was separated into two half leaves, one half-leaf was treated with the test compound, while the other half-leaf was treated with the buffer as the mock control. Finally, the viral fluorescence was imaged using an in vivo imaging system and analyzed using ImageJ software.

3.2 Plant Growth and Maintenance

1. Sown the *N. benthamiana* seeds in pots in a soil mixture (soil: vermiculite = 3:1; v/v).
2. Maintain the pots in a greenhouse under a 16-h light/8-h dark photoperiod with light intensity of 12,000 Lux, 50% relative humidity, and a constant temperature of 26 °C.
3. Transplant *N. benthamiana* seedlings into individual pots at 2 weeks after sowing.
4. Maintain *N. benthamiana* seedlings in a greenhouse under the same conditions and fertilize weekly.
5. Select 4-week-old *N. benthamiana* seedlings with similar size for subsequent experiments.

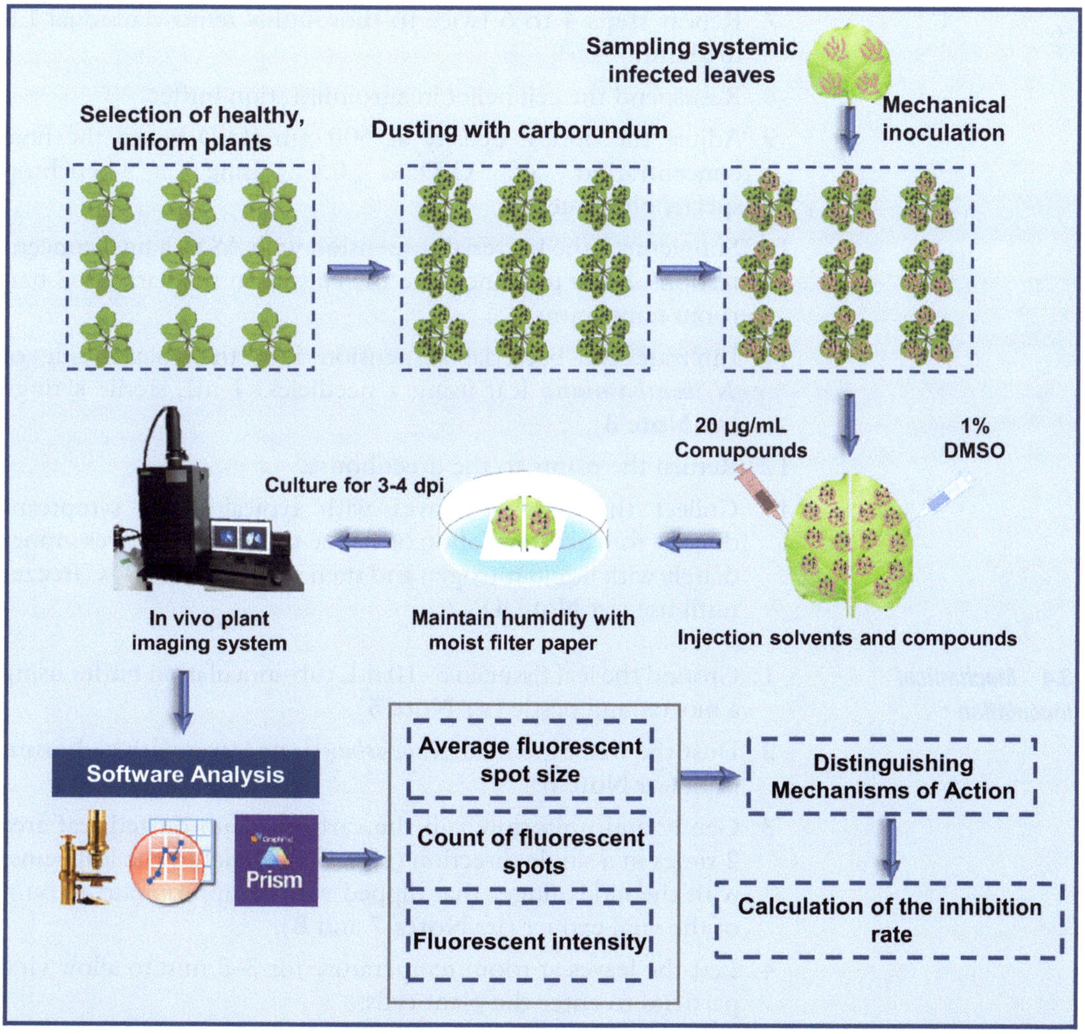

Fig. 1 Technical flowchart

3.3 Preparation of Virus Inoculum

1. Inoculate the liquid LB medium using a sterile pipette tip (for fresh colony) or a pipette (for glycerol stock) of *Agrobacterium rhizogenes* harboring the infectious clone of TMV-GFP (*see* **Note 2**).
2. Place the culture tube in a 28 °C constant temperature shaker incubator and incubate with shaking at 180 rpm for 16–18 h.
3. Transfer the bacterial culture into a sterile 1.5-mL microcentrifuge tube.
4. Centrifuge at 6000 rpm for 1 min at room temperature.
5. Discard the supernatant carefully and add 1 mL of agroinfiltration buffer to the cell pellet.
6. Resuspend the pellet gently with a 1-mL pipette.

7. Repeat **steps 4** to **6** twice to thoroughly remove residual LB medium.
8. Resuspend the cell pellet in agroinfiltration buffer.
9. Adjust the optical density at 600 nm ($O.D._{600}$) to the final concentration of $O.D._{600}$ 0.4 using a benchtop spectrophotometer.
10. Supplement the bacterial suspension with AS to a final concentration of 100 μM. Incubate the culture in the dark for 3 h at room temperature.
11. Infiltrate the bacterial suspension into the abaxial side of *N. benthamiana* leaf using a needleless 1-mL sterile syringe (*see* **Note 3**).
12. Return the plants to the greenhouse.
13. Collect the systemic leaves with typical TMV symptoms directly for rub inoculation or freeze the systemic leaves immediately with liquid nitrogen and then stored in a −80 °C freezer until use (*see* **Note 4**).

3.4 Mechanical Inoculation

1. Ground the leaf tissue in 5–10 mL rub-inoculation buffer using a mortar and pestle (*see* **Note 5**).
2. Dust the well expended *N. benthamiana* leaves with carborundum (*see* **Note 6**).
3. Gently and uniformly rub the carborundum-dusted leaf area 2 times in a single direction (typically parallel to the leaf veins) with the index finger that dipped with an appropriate amount of the viral extract (*see* **Notes 7** and **8**).
4. Left the leaves at room temperature for 3–5 min to allow viral particles to enter the plant cells.
5. Thoroughly rinse the leaf surface with a wash bottle filled with distilled water to remove carborundum particles.
6. Carefully place a prewet filter paper on the inoculated leaf to maintain humidity for 2 h.

3.5 Compound Treatment

1. Dissolving 1 mg of the test chemical in 1 mL of DMSO or distilled water into 1 mg/mL stock solution. Aliquot the stock solution into 100 μL servings and stored at −20 °C (*see* **Note 9**).
2. Dilute the test chemical concentration into 20 μg/mL working solution with distilled water.
3. Infiltrate one half-leaf of the rub-inoculated leaves with 100 μL the control solution contains the same concentration of DMSO using a needleless syringe, while the other half-leaf with the same amount of chemical working solution (*see* **Note 10**).
4. Excise the leaves and immediately bisect longitudinally along the middle vein to avoid potential movement of the chemical between the two half-leaves.

5. Place the treated leaves in Petri dishes with prewet filter paper and seal with Parafilm to keep moisture (*see* **Note 11**).
6. Place the Petri dishes in a plant growth chamber for at least 3 days.

3.6 Fluorescence Imaging

1. Image the fluorescence of the local lesions with a Lumazone Pylon 2048B FL/Double, with the 300-W xenon light source (MAX-303) turned on.
2. Launch the Lumazone LF software, navigate to Plugins and select the Lumazone LF DBoxFL Imaging Tools.
3. Adjust the fluorescence parameters as follows: set the fluorescence light intensity to 100, the fluorescence filter wheel position to 2 (corresponding to 470 nm excitation), and the camera filter wheel position to 2 (corresponding to 525 nm emission).
4. Absorb any water on the leaf samples with a paper towel and place the leaves in the center of the dark chamber with their abaxial side upward.
5. Use the Start Fluorescence Imaging button to capture the fluorescence image and the Start Light Field Imaging button to acquire the bright-field image.
6. Store all photographs in desired fold.

3.7 Image Analysis

1. Open the target image in ImageJ software (ij154-win-java8) [13–15].
2. Use the straight-line selection tool to draw a line corresponding to the known length of the leaf. Use the Set Scale function in the Analyze module to enter the known length of the leaf in the "Distance in pixels" field and set the "Unit of length" (generally center meter).
3. Set the threshold: Select Image → Adjust → Threshold and adjust the two sliders in the window until the target fluorescent lesion appears red against a black background. Minimize the inclusion of background red pixels (noise). Click "Apply" to apply the threshold.
4. Use the Set Measurements function in the Analyze module. In the pop-up dialog box, the following parameters are analyzed: Area, Mean, IntDen.
5. Select the half-leaf area using the rectangular selection tool. Use the Analyze Particles function in the Analyze module. In the "Size" field, enter the minimum and maximum area range from 20 to infinity for the particle analysis. Use the "Display results" and "Summarize" options to display the results for the number of fluorescent foci and the total fluorescence area (*see* **Note 12**).

3.8 Expected Results

The known antiviral agent, namely niningmycin (NNM), was used to illustrate the results of FLLIA. Imaging under ultraviolet revealed an apparent reduction of fluorescent lesion size after NNM treatment compared with mock treatment (Fig. 2a), confirming the antiviral activity of NNM. Statistical analyses further revealed that NNM treatment also reduced the mean fluorescence intensity of TMV-GFP infection lesions (Fig. 2b and c). In addition, the number of fluorescent TMV-GFP lesion in the NNM-treated half-leaf was also significantly less than mock-treated half leaf (Fig. 2d). Collectively, these results suggest NNM can simultaneously target both viral replication and cell-to-cell movement.

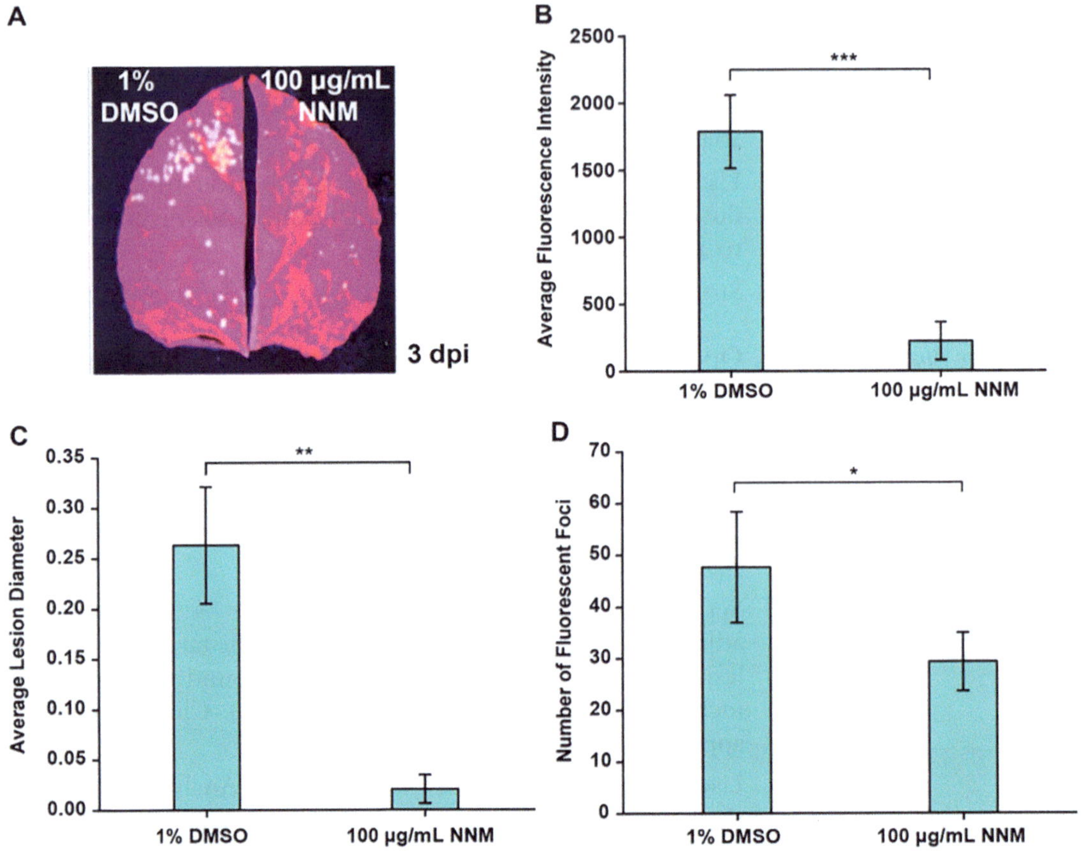

Fig. 2 Representative results of FLLIA. (**a**) Symptoms on leaves at 3 dpi with TMV-GFP. (**b**) Significance analysis of average fluorescence intensity. (**c**) Significance analysis of average diameter. (**d**) Significance analysis of number of fluorescent infection foci. Statistical significance is indicated with asterisks: *** for highly significant, ** for significant, and * for moderately significant differences

4 Notes

1. A recombinant viral infectious clone expressing a free fluorescent protein (e.g., GFP, YFP, or mCherry) is necessary for FLLIA to facilitate visual tracking of viral infection.
2. To preserve genetic stability and biological activity, the original bacteria harboring viral infectious clone can be aliquoted in 15% glycerol, strictly avoiding repeated freeze–thaw cycles.
3. For rub-inoculation, select fully expanded, mature leaves from the middle part of the plant, which are metabolically active and developmentally stable, favoring reproducible outcomes.
4. Virus-infected leaves can be wrapped in aluminum foil before being frozen in liquid nitrogen.
5. Before grinding, pre-chill the mortar and pestle.
6. Use minimize carborundum to avoid mechanical damage.
7. During rub-inoculation with the plant viral extract, apply minimal pressure to minimize damage to the leaf tissue and ensure rubbing is performed in a single direction.
8. To ensure experimental accuracy, inoculate the left and right half-leaves of the same leaf with equal amounts of viral extract, and maintain consistent rubbing pressure and area across replicates.
9. Change gloves when handling different compounds to prevent cross-contamination.
10. Perform multiple localized injections across the leaf surface until the solution infiltrates the majority of the half-leaf area.
11. Include no fewer than three leaves of uniform size and physiological state as independent biological replicates for each compound treatment to ensure statistical power.
12. Maximum area range should be adjust based on the lesion size.

References

1. Yu L, Korxeelor X, Wang Z et al (2025) The 6-kilodalton peptide 1 of the family *Potyviridae*: small in size but powerful in function. Front Microbiol 16:1605199
2. Chai M, Li L, Li Y et al (2024) The 6-kilodalton peptide 1 in plant viruses of the family *Potyviridae* is a viroporin. PNAS 121(21):e2401748121
3. Bradamante G, Mittelsten Scheid O, Incarbone M (2021) Under siege: virus control in plant meristems and progeny. Plant Cell 33(8): 2523–2537
4. Fang L, Geng C, Wei X-Y et al (2024) Potato virus Y viral protein 6K1 inhibits the interaction between defense proteins during virus infection. Plant Physiol 194(3):1447–1466
5. Zan N, Li J, Yao J et al (2025) Rational design of *phytovirucide* inhibiting nucleocapsid protein aggregation in tomato spotted wilt virus. Nat Commun 16(1):2034
6. Hu W, Deng C, Qin L et al (2025) A conserved lysine/arginine-rich motif is essential for the autophagic degradation of *potyviral* 6K1 protein and virus infection. J Virol 99(3): e0218324

7. Hýsková V, Bělonožníková K, Chmelík J et al (2024) *Potyviral* helper-component protease: multifaced functions and interactions with host proteins. Plants 13(9):1236
8. Xue Q, Swevers L, Taning CNT (2023) Plant and insect virus-like particles: emerging nanoparticles for agricultural pest management. Pest Manag Sci 79(9):2975–2991
9. Tatineni S, Hein GL (2023) Plant viruses of agricultural importance: current and future perspectives of virus disease management strategies. Phytopathology 113(2):117–141
10. Tsai WA, Brosnan CA, Mitter N et al (2022) Perspectives on plant virus diseases in a climate change scenario of elevated temperatures. Stress Biol 2(1):37
11. Yan Y, Huang M, Wang L et al (2022) Insights into a rapid screening method for anti-cucumber mosaic virus compounds. J Virol Methods 301:114402
12. Wang L, Tan H, Medina-Puche L et al (2022) Combinatorial interactions between viral proteins expand the potential functional landscape of the tomato yellow leaf curl virus proteome. PLoS Pathog 18(10):e1010909
13. Schneider CA, Rasband WS, Eliceiri KW (2012) NIH image to ImageJ: 25 years of image analysis. Nat Methods 9(7):671–675
14. Rueden CT, Schindelin J, Hiner MC et al (2017) ImageJ2: ImageJ for the next generation of scientific image data. BMC Bioinformatics 18(1):529
15. Schindelin J, Arganda-Carreras I, Frise E et al (2012) Fiji: an open-source platform for biological-image analysis. Nat Methods 9(7): 676–682

Chapter 17

Enhancing Plant Antiviral Resistance with Microbial Immunomodulators: A Half-Leaf Spot Screening Method

Chao Ma, Kelei Han, and Dankan Yan

Abstract

The hypersensitive response (HR) represents a critical antiviral pathway in plants that can be leveraged for therapeutic screening via half-leaf spot assay. Attenuation of HR-induced necrotic lesions serves as a quantitative readout of antiviral efficacy. Here we present an integrated protocol combining: (1) rhizosphere bacteria isolation from asymptomatic plants in virus-infecting fields via serial dilution plating; and (2) quantitative antiviral assessment through necrotic lesion analysis. This comprehensive approach identifies microbial immunomodulators with potential for sustainable plant virus control.

Key words Plant resistance, Plant virus, Immunomodulators, Half-leaf spot method

1 Introduction

Plant viruses (over 2000 species) cause significant agricultural losses due to global spread, latency, and high mutation rates [1]. Conventional pesticides are often ineffective and environmentally harmful, necessitating sustainable alternatives like microbial biopesticides. However, few specific antiviral agents are currently available for the control of plant viral diseases. *Pseudomonas oleovorans* strain KBPF-004 has been shown to reduce seed transmission of cucumber green mottle mosaic virus (CMV) and pepper mild mottle virus (PMMoV) [2]. Similarly, cellulases have been reported to suppress viral infection in plants [3].

The half-leaf spot assay is a standard technique for quantifying viral infectivity and screening antiviral compounds, wherein inoculation of *N. tabacum* Samsun leaves with viral extracts produces necrotic lesions that serve as a quantitative readout of viral load [4]. Recent demonstrations of benzimidazole derivatives' efficacy against PVY highlight the assay's versatility for evaluating both chemical and microbial agents.

Guanwei Wu and Xiaofei Cheng (eds.), *Plant Antiviral Resistance: Methods and Protocols*, Methods in Molecular Biology, vol. 3039, https://doi.org/10.1007/978-1-0716-5300-5_17,

Here, we present a detailed protocol employing this method to screen soil bacteria for antiviral activity using PMMoV as a case, providing a robust framework for identifying microbial immunomodulators toward sustainable plant virus management.

2 Materials

2.1 Buffers and Reagents

1. Nutrient Agar (NA) medium: dissolve 3 g beef extract, 10 g peptone, 5 g NaCl, and 15 g agar in distilled water to a final volume of 1 L (for liquid NA medium: omit agar). Adjust pH to 7.5. Autoclave at 121 °C for 15 min.
2. 25% (v/v) glycerol supplemented NA medium: dissolve 3 g beef extract, 10 g peptone, 5 g NaCl, and 250 mL glycerol in distilled water, then adjust to a final volume of 1 L, and autoclave at 121 °C for 15 min.
3. Physiological saline solution: dissolve 0.9 g NaCl in distilled water to a final volume of 100 mL, and autoclave at 121 °C for 15 min.
4. PBS buffer: dissolve 8.0 g NaCl, 0.2 g KCl, 1.44 g Na_2HPO_4, and 0.24 g KH_2PO_4 in distilled water to a final volume of 1 L and sterilize by autoclaving, adjust pH to 7.4.
5. Ningnanmycin working solution: dilute 8% Ningnanmycin stock 1:800 in sterile water.

2.2 Plant Materials

Germinate *N. tabacum* Samsun seeds in seedling trays containing sterile potting soil under controlled conditions (23–26 °C, 16 h of light per day, 60% humidity). After 7–10 days, transplant seedlings into individual pots and grown for 14 additional days until plants reach the 4–6 leaf stage.

2.3 Virus Inoculum

PMMoV-infected plant tissues are kept in Laboratory of Plant Virology at Anhui Academy of Agricultural Sciences.

2.4 Bacterial Strains

Pseudomonas sp. stain ZTT7-4 are kept in Laboratory of Plant Virology at Anhui Academy of Agricultural Sciences.

3 Methods

3.1 Soil Sampling and Bacterial Isolation

1. Collect Rhizosphere soil at 10 cm depth from healthy or lightly symptomatic plants in a field exhibiting severe viral disease (stunted growth, leaf mottling, chlorosis, and deformation). Obtain samples using sterile tools, stored in sterile bags, and transport them on ice.

2. Weigh 5 g of soil sample and add 50 mL of sterile physiological saline solution. Mix thoroughly with glass beads for 10 min, then let it stand for 10 min to allow sedimentation.
3. Aspirate the supernatant and dilute it at a ratio of 1:10 (e.g., 1 mL of the original solution with 9 mL of physiological saline) to prepare a 10^{-1} dilution. Continue with ten-fold serial dilutions (10^{-2}, 10^{-3}, etc.) for future use. Pipette 100 μL of the diluent (e.g., 10^{-3} dilution) and evenly spread it on the surface of a nutrient agar plate using a sterile pipette (*see* **Notes 1** and **2**).
4. Gently tap the plate surface with a sterile glass bead to ensure even distribution of the bacterial suspension. Invert the plate and incubate it at 30 °C in incubator for 24–48 h. Select distinct colonies exhibiting defined morphotypes. Transfer isolates with smooth margins and $\geq$ 2 mm diameter aseptically using sterile inoculation loops to NA medium plates (*see* **Note 3**).
5. Subculture inverted-incubated plates 2–3 times until homogeneous pure cultures are obtained. Preserve resulting isolates in 25% (v/v) glycerol supplemented NA medium at −80 °C or utilize as test inoculum for half-leaf spot method assays.

3.2 Antiviral Activity Assay

1. Using ZTT7-4 as a case, grow it in liquid NA medium at 28 °C and 200 rpm until OD_{600} reaches 0.8 (*see* **Note 4**).
2. Select healthy, uniform *N. tabacum* Samsun plants at the 4–6 leaf stage for antiviral activity assay (*see* **Note 5**).
3. Apply ZTT7-4 suspension evenly to the right half of each leaf (treatment, T) and an equal volume of NA liquid medium to the left half (control, C).
4. After 12 h at 25 °C, uniformly challenge all leaves with PMMoV inoculum (1 mg/mL in PBS) using silicon dioxide. Incubate plants at 25 °C, 60% RH, 16/8 h light/dark for 48 h (*see* Fig. 1a; **Notes 6–9**). Representative results are shown in Fig. 1a.
5. Calculate the preventive efficacy using the formula: Inhibition rate (%) = $[(C - T)/C] \times 100$, where C and T represent lesion counts on control group and treatment group leaf halves.
6. Conduct concurrent quality control experiments: coat right leaf halves with ningnanmycin (treatment, T) and left halves with sterile distilled water (SDW) as described above (Fig. 1b). Validate results only if treatment inhibition rate exceeds 50%.

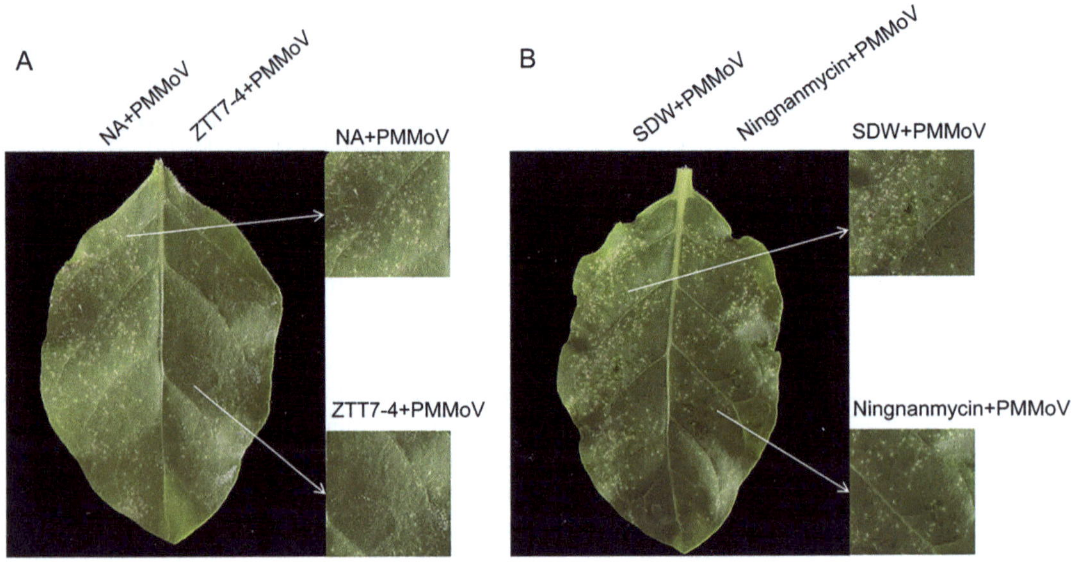

Fig. 1 Screening of bacterial antiviral agents against PMMoV infection in *N. tabacum* Samsun by half-leaf spot assay. (**a**) Left half (control): liquid NA medium + PMMoV; Right half (treatment): ZTT7-4 + PMMoV. (**b**) Quality control. Left half (control): sterile distilled water (SDW) + PMMoV; Right half (treatment): ningnanmycin + PMMoV

4 Notes

1. Maintain sterile technique during soil bacterial isolation by working near an alcohol burner in a laminar flow hood. Pre-test media sterility by incubating at 37 °C for 24 h before use.
2. Dilute soil suspension to 10^{-6}–10^{-7} to obtain well-isolated single colonies.
3. Select and preserve candidate strains based on distinct colony morphology (edges, elevation, color, pigmentation, lytic zones) to facilitate subsequent species identification.
4. It is important to be aware that different strains may require varying incubation times and temperatures. Prior to initiating the experiment, it is advisable to perform preliminary shake flask cultures from individual colonies on plates to ascertain the optimal culture conditions. The OD_{600} of the bacterial culture for experimental use should be greater than 0.8 but should not exceed 1.0.
5. Maintain strict fertilization, irrigation, and pest control (particularly aphids) during plant cultivation to prevent adventitious viral infection.
6. Use 70–110 mesh silicon carbide for mechanical inoculation of *N. tabacum* Samsun. Apply optimal pressure to generate a

countable number of necrotic lesions for accurate quantification.

7. For PMMoV inoculation, RT-PCR or Western blot can be employed for preliminary detection. The virus source can be stored at −80 °C, but it is recommended to use fresh leaves as the inoculum for the half-leaf spot method assay.
8. Use a watercolor brush for high-throughput inoculation of uniform material. Rinse inoculated leaves thoroughly after inoculation to enhance infection efficiency.
9. Sterilize all inoculation tools by boiling (100 °C, 10 min) and wash hands thoroughly with soap to prevent cross-contamination. Steam-sterilize pots and soil when possible.

Acknowledgments

This work was financially supported by the National Natural Science Foundation of China (32300150), the Youth Development Fund from Anhui Academy of Agricultural Sciences (QNYC-202506, QNYC-202507), the National Key Research and Development Program of China (2023YFD1401000) and Key Laboratory of Biotechnology in Plant Protection, Ministry of Agriculture and Rural Affairs and Zhejiang Province; Key Laboratory of Green Plant Protection, Zhejiang Province (NYBQS-KF2025002).

References

1. Wu J, Zhang Y, Li F et al (2024) Plant virology in the 21st century in China: recent advances and future directions. J Integr Plant Biol 66: 579–622
2. Kim NG, Seo EY, Han SH et al (2017) *Pseudomonas oleovorans* Strain KBPF-004 culture supernatants reduced seed transmission of cucumber green mottle mosaic virus and pepper mild mottle virus, and remodeled aggregation of 126 kDa and subcellular localization of movement protein of pepper mild mottle virus. Plant Pathol J 33:393–401
3. Oka N, Ohki T, Honda Y et al (2008) Inhibition of pepper mild mottle virus with commercial cellulases. J Phytopathol 156:65–67
4. Holmes FO (1954) Inheritance of resistance to viral diseases in plants. Adv Virus Res 2:1–30

Chapter 18

Screening and Identification of Antiviral Chemical Components from Medical Plants

Chenggang Song, Ruifang Gao, Xinyun Cui, and Jianchun Qin

Abstract

Medicinal plants constitute a crucial reservoir of natural antiviral compounds. Efficient isolation and molecular structure determination are essential for advancing their utilization. This chapter delineates principal methods for purifying antiviral compounds from medicinal plants and elucidating unknown molecular structures, supplemented with representative research examples. Additionally, a general methodology for evaluating antiviral activity against plant viruses is presented to expedite compound identification and purification. The integrated approaches and insights provide a valuable framework for the discovery of natural antiviral compounds from botanical sources.

Key words Medicinal plant, Natural products, Antiviral, Extraction and isolation, Identification, Activity test

1 Introduction

Plant viral diseases are significant threats to global agricultural production, causing over 10% of crop yield losses annually and severely undermining the stability of food security systems [1]. These diseases are primarily transmitted through insect vectors, mechanical contact, or seeds, and they infect a wide range of hosts, including cash crops, ornamental plants, and forest resources [2]. Plant viruses accelerate their spread within host plants by damaging the integrity of cell membrane structure and interfering with the normal physiological metabolism of plants [3]. Typical viral pathogens include tobacco mosaic virus (TMV) and cucumber mosaic virus (CMV), which often lead to leaf yellowing and curling, fruit malformation, and even complete crop failure in severe cases [4]. While current major control strategies focus on the breeding of disease-resistant varieties and the application of biological pesticides [5, 6], the rapid mutation rates of viruses typically render

Guanwei Wu and Xiaofei Cheng (eds.), *Plant Antiviral Resistance: Methods and Protocols*, Methods in Molecular Biology, vol. 3039, https://doi.org/10.1007/978-1-0716-5300-5_18,

single-method approaches inadequate for comprehensive prevention.

Traditional control methods predominantly rely on chemical pesticides. However, long-term use of these pesticides tends to induce viral resistance and cause environmental pollution [7]. In this context, the exploration and application of natural antiviral components offer a new development direction for the green control of plant viral diseases [8]. In nature, many organisms possess defense systems against pathogens through their own metabolic products, providing new ideas for discovering antiviral ingredients. For example, lentinan is a broad-spectrum biological agent for plant viral diseases and a biological pesticide extracted from mushroom culture media. Lentinan can directly inactivate TMV virions in vitro, rendering them incapable of infection. It also inhibits the proliferation of viral RNA in plants, reduces the amount of viral RNA, and simultaneously suppresses the replication of viral RNA and the synthesis of viral proteins to achieve virus control [9, 10]. Ningnanmycin, discovered from *Streptomyces*, can effectively prevent and control TMV in crops and has been commercially applied on a large-scale in China [11]. Nevertheless, long-term use of a single pesticide will enhance viral resistance and reduce the efficacy of the pesticide [12]. Therefore, the development of more new antiviral active components remains a top priority in the prevention and control of plant viral diseases.

Medicinal plants serve as important sources of natural antiviral components, and various active constituents derived from these plants have demonstrated remarkable antiviral efficacy. For instance, matrine-type components isolated from *Sophora tonkinensis* (*Vietnamese sophora*) and quinolizidine alkaloid components identified from *Thermopsis lanceolata* (lance-leaved thermopsis) have shown significant inhibitory effects on plant viruses such as TMV and tomato spotted wilt virus (TSWV) [13, 14]. The discovery and identification of these active constituents have established a crucial theoretical foundation for the development of targeted antiviral agents.

The effective utilization of natural antiviral components relies on precise extraction, separation, and identification. Current isolation protocols primarily combine diverse extraction techniques with column chromatography, employing activity-guided fractionation to enhance separation efficiency. Structural elucidation of compounds integrates mass spectrometry, nuclear magnetic resonance spectroscopy, and other spectroscopic techniques to achieve accurate structural elucidation. This chapter systematically introduces and elaborates on the extraction, isolation, purification, structural characterization, and activity evaluation of natural antiviral components derived from medicinal plants, aiming to provide a theoretical reference for the exploration and application research of natural antiviral agents.

2 Materials

2.1 Biological Materials

1. Target medicinal plant (e.g., the whole plant of *Dendrobium findlayanum* collected from Mengla, Yunnan Province).
2. *Nicotiana benthamiana* seeds (Laboratory of Plant Pathology at College of Plant Science, Jilin University).
3. TMV strain was obtained from College of Biological Sciences, China Agricultural University.

2.2 Chemicals

1. Ethanol (C_2H_5OH).
2. Methanol (CH_3OH).
3. Acetone (CH_3COCH_3).
4. Petroleum ether.
5. Ethyl acetate ($CH_3COOC_2H_5$).
6. Dichloroethane (CH_2Cl_2).
7. *n*-Butanol ($CH_3(CH_2)_3OH$).
8. Chloroform-*d* ($CDCl_3$).
9. Dimethyl sulfoxide-d_6 (C_2D_6SO).
10. Silica gel.
11. TransStart Top Green qPCR SuperMix.

2.3 Equipment

1. Rotary evaporator.
2. Mass spectrometer.
3. Nuclear magnetic resonance spectrometer.
4. Infrared spectrometer.
5. Chirascan spectrometer.
6. Quantitative Real-Time PCR instrument.

3 Methods

The process of isolation and characterization of antiviral constituents from medicinal plants proceed through the sequential extraction, separation, purification, structural identification, and activity evaluation (Fig. 1).

3.1 Extraction of Antiviral Components from Medicinal Plants

1. Crush dried whole-plant material, soak in methanol at room temperature for 1 h, sonicate in a 50 °C water bath for 20 min, repeat extraction 3 times, and remove residue.
2. Concentrate the methanol extract under vacuum at 45 °C using a rotary evaporator. Partition the crude extract

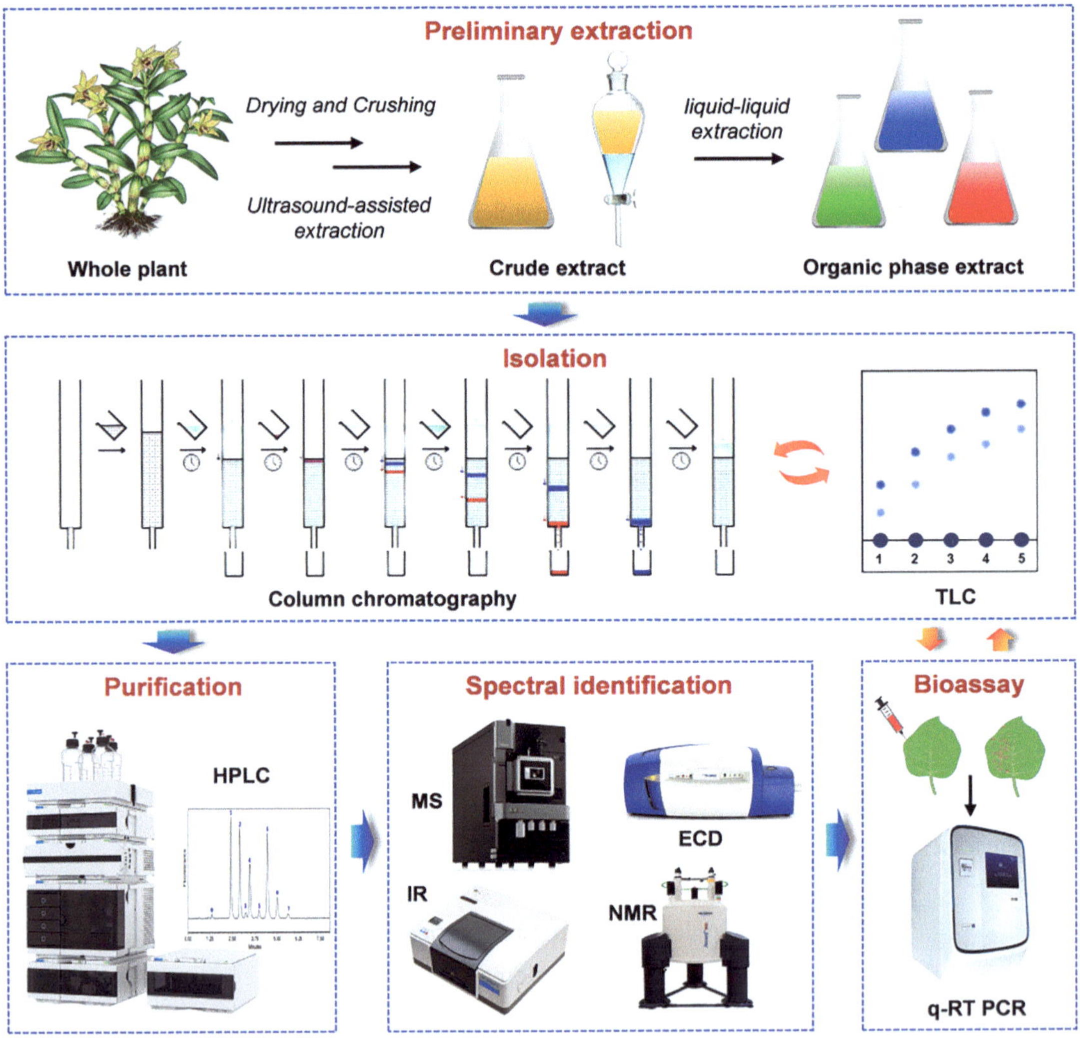

Fig. 1 Workflow for obtaining antiviral active ingredients from medicinal plants

sequentially with petroleum ether, ethyl acetate, and n-butanol in a separatory funnel.

3. Evaporate the four fractions (petroleum ether, ethyl acetate, n-butanol, and aqueous phase) to dryness and store at 4 °C.

3.2 Separation and Purification of Antiviral Components

3.2.1 Preliminary Separation Using Column Chromatography

1. Pack silica gel into a glass column, equilibrate with low-polarity solvent (e.g., petroleum ether/dichloroethane), load sample, and elute to separate components by polarity.
2. Increase eluent polarity gradually (e.g., add acetone/methanol) and collect fractions sequentially.
3. Perform gel column chromatography (e.g., Sephadex LH-20): pack gel, equilibrate with methanol or water–methanol

mixture, load silica gel fractions, and elute with the same solvent (*see* **Note 1**).

4. Monitor fractions by thin-layer chromatography (TLC), combine similar fractions, concentrate, and dry to obtain preliminary active components.

3.2.2 Tracking Target Fractions Using Thin-Layer Chromatography

1. Draw a straight line with a pencil 1.5–2.0 cm from the bottom edge of the TLC plate as the spotting baseline. Spot sample solution on baseline using a capillary tube, with spot diameter controlled at 2–3 mm and inter-spot distance maintained at 0.5–1.5 cm to prevent cross-contamination.
2. Pour a suitable eluent into the developing chamber to a depth of 0.5–1 cm, and place the spotted plate inside, ensuring the baseline remained above the solvent level before sealing the chamber. Remove the plate immediately when the solvent front reaches approximately 1 cm from the top (about 4/5 of the plate length), and promptly mark the solvent with a pencil.
3. After drying, visualize and mark spots under UV light at 254 or 365 nm.
4. Calculate the retardation factor (Rf value) as Rf = a/b, where "a" represents the distance from the origin to the spot center, and "b" represents the distance from the origin to the solvent front (Fig. 2). Under identical experimental conditions, the sample and reference standard are spotted in parallel on the same TLC plate. If a sample spot exhibits identical Rf value and color to the reference, it is preliminarily identified as the same compound. A single spot indicates high purity, while multiple spots suggest a mixture.

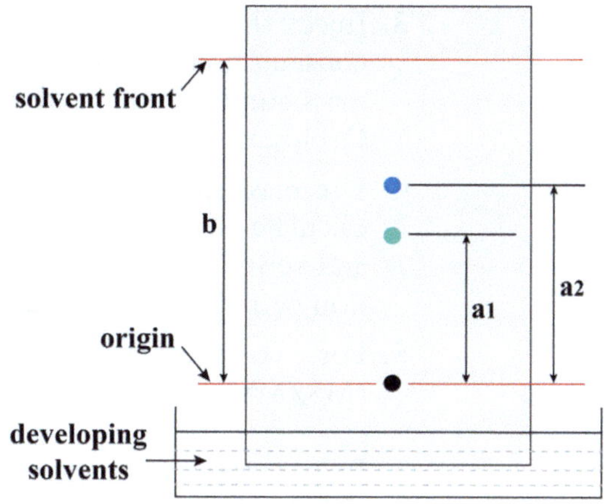

Fig. 2 Schematic diagram of thin-layer chromatography separation principle

3.2.3 Purification of Target Compounds Using High-Performance Liquid Chromatography (HPLC)

1. Before loading, dissolve the sample to be purified in methanol water (1:1), filter it through a 0.45-μm microporous membrane, and control the loading amount at 5–10% of the chromatographic column load to avoid overload.
2. Select a C18 reverse phase column (such as 250 mm × 4.6 mm, particle size 5 μm) as the chromatographic column, use purified water as the A phase and acetonitrile/methanol as the B phase eluent as the mobile phase, and set different gradient elution programs according to the polarity of the fractions, such as increasing the acetonitrile ratio from 20% to 40% in 0–15 min.
3. Maintain a flow rate of 1.0 mL/min, column temperature of 25 °C, and set the detection wavelength and detection wavelength to 254 and 210 nm, respectively.
4. Concentrate target peak fraction using rotary evaporator at 40 °C. Freeze-dry to obtain purified product.

3.3 Structural Elucidation of Antiviral Constituents from Medicinal Plants

Structural characterization is essential for elucidating a compound's mechanism of action. The purified compound is identified as follows:

3.3.1 Molecular Weight Testing and Formula Calculation

1. Dissolve HPLC-purified compound in methanol or acetonitrile to prepare 0.1–1 mg/mL solution.
2. Turn on the mass spectrometer and perform system calibration to ensure a quality accuracy of better than 5 ppm. Use electro spray ionization (ESI) or atmospheric pressure chemical ionization (APCI) sources to set positive and negative ion modes, and select appropriate ionization modes according to the properties of the compounds.
3. Inject the sample solution into the mass spectrometer at a constant flow rate (such as 5–10 μL/min), and obtain accurate mass numbers through high-resolution mass analyzers (such as Orbitrap or TOF).
4. Use mass spectrometry workstation software to process data, calculate element composition through precise mass numbers, and verify molecular formulas by combining isotope distribution patterns (*see* **Note 2**).
5. For complex samples, multi-stage mass spectrometry (MS/MS) can be used for structural analysis.

3.3.2 Detection of Functional Groups

1. Prepare the sample in one of three forms: potassium bromide (KBr) pellet, liquid film, or solid pressed into a thin layer. For KBr pellet preparation, mix 1–2 mg of dried sample with 100–200 mg of dried KBr powder, grind thoroughly, and press into a transparent pellet under high pressure.

2. Place the sample in the sample holder of the infrared spectrometer. Set the instrument parameters: scan range 4000–400 cm^{-1}, resolution 4 cm^{-1}, and accumulate 32 scans per sample.
3. Analyze the spectrum by identifying characteristic absorption peaks: O-H stretching (3200–3600 cm^{-1}), C–H stretching (2850–3000 cm^{-1}), C=O stretching (1600–1800 cm^{-1}), and fingerprint region (400–1500 cm^{-1}).
4. Compare the obtained spectrum with standard reference spectra or spectral databases to identify specific functional groups.

3.3.3 Analysis of the Planar Structure of Compounds by Nuclear Magnetic Resonance (NMR) Spectroscopy

1. Dissolve 5–10 mg sample in 0.5 mL deuterated solvent (CDCl3 or DMSO-d6), and transfer to 5 mm NMR tube.
2. Analyze the sample using a high-resolution NMR spectrometer operating at 400 MHz for ^{1}H NMR and 100 MHz for ^{13}C NMR. The ^{1}H NMR spectrum is acquired with a relaxation delay of 1.5 s, acquisition time of 2.5 s, and 32 scans. The ^{13}C NMR spectrum is obtained using inverse gated decoupling with 2048 data points and 5000 scans.
3. Reference chemical shifts to residual solvent peaks ($CDCl_3$: δ_H 7.26, δ_C 77.16; DMSO-d_6: δ_H 2.50, δ_C 39.52). Process data using standard NMR software. Analyze coupling patterns, integration values, and chemical shifts to deduce planar structure (*see* **Note 3**).
4. Compare the resulting spectra with literature data for structural confirmation.

3.3.4 Absolute Configuration Analysis of Compounds by Electronic Circular Dichroism (ECD) Analysis

1. Dissolve the purified compound in HPLC-grade methanol or ethanol to prepare a 0.1–0.5 mM solution.
2. Place the sample in a quartz cuvette with a 1-mm path length and analyze using a high-resolution CD spectrometer operating in the 200–400 nm range. Acquire the ECD spectrum with a 1 nm bandwidth, 1 nm data pitch, and 2 s response time per point, averaging 3 scans to enhance signal-to-noise ratio. The instrument is calibrated with d-10-camphorsulfonic acid before measurement.
3. Analyze the resulting Cotton effect patterns using time-dependent density functional theory (TD-DFT) calculations to predict ECD spectra for possible stereoisomers.
4. Compare the experimental spectrum with theoretical predictions to assign the absolute configuration (*see* **Note 4**).

3.4 Evaluation of Antiviral Activity of Phytochemicals from Medicinal Plants

3.4.1 Local Lesion Count Assay (Half-Leaf Method)

1. Grow tobacco seedlings to the 5–6 leaf stage for subsequent use.
2. Prepare viral inoculum by homogenizing 1 g of TMV-infected *Nicotiana tabacum* leaves in 1 mL of 0.1 M phosphate buffer (pH 7.1).
3. Mechanically inoculate virus-containing sap onto the tobacco leaves by friction with carborundum as an abrasive.
4. At 24 h post-inoculation, treat the upper right half of each leaf with plant-derived compounds; the upper left half serves as the untreated control.
5. Assess antiviral activity by counting local lesions and calculate percentage inhibition using the following formula:

$$[I(\%) = (1 - T/C) \times 100]$$

   where T and C represent the number of local lesions on the treated and untreated leaf halves, respectively.

3.4.2 Quantitative Real-Time PCR (qRT-PCR) Analysis

1. Extract total RNA from treated plant tissues using a commercial kit.
2. Reverse-transcribe RNA into first-strand complementary DNA (cDNA) using reverse transcriptase and random primers.
3. Amplify a segment of the messenger RNA (mRNA) sequence of elongation factor 1-alpha (EF1α) as an internal reference control using specific primers during qRT-PCR.
4. Analyze the qRT-PCR data using the $2^{-\Delta\Delta Ct}$ method.

4 Notes

1. Throughout the process, attention must be paid to solvent selection, column equilibration, and elution gradient control to ensure effective separation.
2. During the mass experimental process, attention should be paid to sample purity, solvent selection, and optimization of ion source parameters to ensure data accuracy.
3. Ensure the sample is dry and free from moisture contamination for accurate IR results.
4. For chiral compounds with multiple stereocenters, complementary methods like X-ray crystallography or nuclear overhauser effect (NOE) experiments were used to confirm the assigned configuration.

Acknowledgments

This work is funded by the National Natural Science Foundation of China (32201781), the Science and Technology Development Plan Project of Jilin Province (20230202019NC), the eighth Jilin Province Young Sci-Tech Talents Cultivation Project (QT202423), and the tenth Jilin University Interdisciplinary Cultivation Program for Young Faculty and Students (2024-JCXK-17).

References

1. Tatineni S, Hein GL (2023) Plant viruses of agricultural importance: current and future perspectives of virus disease management strategies. Phytopathology 113:117–141
2. Wang A (2015) Dissecting the molecular network of virus-plant interactions: the complex roles of host factors. Annu Rev Phytopathol 53:45–66
3. Beatriz Moreno A, Jose Lopez-Moya J (2020) When viruses play team sports: mixed infections in plants. Phytopathology 110:29–48
4. Su CF, Das D, Aslam MM, Xie JQ, Li XY, Chen MX (2023) Eukaryotic splicing machinery in the plant-viru battleground. Wiley Interdiscip Rev: RNA 14:e1793
5. O'Brien PA (2017) Biological control of plant diseases. Australas Plant Pathol 46:293–304
6. Rubio L, Galipienso L, Ferriol I (2020) Detection of plant viruses and disease management: relevance of genetic diversity and evolution. Front Plant Sci 11:1092
7. Hough J, Howard JD, Brown S, Portwood DE, Kilby PM, Dickman MJ (2022) Strategies for the production of dsRNA biocontrols as alternatives to chemical pesticides. Front Bioeng Biotechnol 10:980592
8. Islam W, Adnan M, Tayyab M, Hussain M, Ul Islam S (2018) Phyto-metabolites; an impregnable shield against plant viruses. Nat Prod Commun 13:105–112
9. Wang J, Wang HY, Xia XM, Li PP, Wang KY (2013) Inhibitory effect of sulfated lentinan and lentinan against tobacco mosaic virus (TMV) in tobacco seedlings. Int J Biol Macromol 61:264–269
10. Wang J, Yu G, Li Y et al (2015) Inhibitory effects of sulfated lentinan with different degree of sulfation against tobacco mosaic virus (TMV) in tobacco seedlings. Pestic Biochem Physiol 122:38–43
11. Li X, Hao G, Wang Q et al (2017) Ningnanmycin inhibits tobacco mosaic virus virulence by binding directly to its coat protein discs. Oncotarget 8:82446–82458
12. Han Y, Luo Y, Qin S et al (2014) Induction of systemic resistance against tobacco mosaic virus by Ningnanmycin in tobacco. Pestic Biochem Physiol 111:14–18
13. Zhang T, Tian CY, Zhang J et al (2024) Quinolizidine alkaloids and isoflavones from the herb of *Thermopsis lupinoides* and their antiviral, antifungal, and insecticidal activities. J Agric Food Chem 72:5047–5061
14. Zhang J, Zhang T, An Q et al (2024) Five quinolizidine alkaloids with anti-tobacco mosaic virus activities from two species of *Sophora*. Chin Chem Lett 35:108927

Chapter 19

Development of a Reverse Genetics System for Rice Tiller Inhibition Virus

Wencheng Liu, Wenkai Yan, and Zhongxin Guo

Abstract

As a globally vital food crop, rice suffers severe yield losses from viral diseases. While over a dozen rice viruses have been documented in China, functional reverse genetics system remains unavailable for most, significantly limiting molecular analysis and breeding for virus resistance. This chapter presents a detailed protocol for establishing a viral reverse genetics system using rice tiller inhibition virus (RTIV) as a case. The methodology provides a robust framework for developing infectious clones of other rice viruses.

Key words Rice tiller inhibition virus, Infectious cDNA clone, Rice

1 Introduction

Rice (*Oryza sativa* L.) is one of the most critical staple crops, feeding over half of the world's population [1], its production stability directly impacts food security and socio-economic development. However, rice cultivation faces significant challenges from biotic stresses, with viral diseases emerging as a major threat due to their rapid spread and potential to cause severe yield losses [2, 3]. In China, the most damaging rice viruses are primarily negative-sense RNA viruses and double-stranded RNA viruses, such as *rice stripe virus* (RSV), *southern rice black-streaked dwarf virus* (SRBSDV), and *rice stripe mosaic virus* (RSMV) [4]. In the fields, these rice viruses are mainly transmitted and infected the cultivated rice through insects, such as leafhoppers or planthoppers, leading to characteristic disease symptoms that include plant dwarfing, yellow stripes, mosaic, which causes severe yield losses [5, 6]. Currently, the reverse genetics system of these rice viruses has not yet been established, which significantly limits detailed molecular genetic analysis of all aspects of viral infection cycle and pathogenicity. Recently, Feng et al. have developed a mini-replicon-based reverse-genetics system for RSV in *Nicotiana benthamiana*, which

Guanwei Wu and Xiaofei Cheng (eds.), *Plant Antiviral Resistance: Methods and Protocols*, Methods in Molecular Biology, vol. 3039, https://doi.org/10.1007/978-1-0716-5300-5_19,

represents the first mini-replicon-based reverse-genetics system for monocot-infecting tenuivirus, but this system still fails to successfully infect rice [7].

The reverse genetics technology system based on infectious clones is the most used approach in plant virology research which has revolutionized plant virology by enabling functional studies of viral replication, transcription, movement, and host factors involved in interactions between the virus and the host plant [8–10]. Since the first construction of an infectious clone for *Tomato golden mosaic virus* (TGMV) in 1982 [11], the application of infectious clones for plant DNA viruses has become widespread. Subsequently, the first successful infectious clone of a plant single-stranded positive-sense RNA virus was achieved in 1984, where the in vitro transcription products of a *Brome mosaic virus* (BMV) cDNA clone were infectious and could systemically infect barley plants [12]. For plant single-stranded negative-sense RNA viruses, the first successful minireplicon of *Sonchus yellow net rhabdovirus* (SYNV) was experimentally demonstrated in 2013 [13], followed by the establishment of a reverse genetics system for this virus in 2015 [14]. The reverse genetics system for plant multipartite negative-sense RNA viruses was first successfully constructed in 2020 [15]. However, the reverse genetics system for plant double-stranded RNA viruses remains unachieved.

Recently, we identified rice tiller inhibition virus (RTIV), a novel aphid-borne positive-strand RNA viruses from wild rice using metagenomic sequencing, ribosomal RNA-depleted transcriptome assembly, and virus-derived small interfering RNAs (vsiRNA) analysis [16–18], highlighting the potential of spillover to cultivated rice. Here, we present a detailed protocol for developing a reverse genetics system for RTIV. The resulting infectious clone successfully infects rice plants and exhibits characteristic symptoms. This RTIV reverse genetics platform provides a robust framework for establishing infectious clones of other rice viruses.

2 Materials

2.1 Plant Materials and Bacterial Strains

1. Seeds of rice (*Oryza sativa* L.) cultivar Zhonghua 11 and various wild rice species.
2. Competent cells of *Escherichia coli* strains DH5α.
3. Competent cells of *Agrobacterium tumefaciens* strain GV3101.

2.2 Plasmids

The empty plasmid pCass4-Rz, containing multicloning sites (MCS), a double CaMV 35S promoter (35S) followed by a ribozyme sequence cassette derived from satellite tobacco ring spot virus (Rz) and a Nos terminator (T) (*see* **Note 1**).

2.3 Reagents and Buffers

1. Rapid amplification of cDNA ends (RACE) kits.
2. The first strand cDNA synthesis kit.
3. DNA polymerase.
4. Taq DNA polymerase.
5. Gel Extraction Kit.
6. Plasmid purification kit.
7. Recombinant DNA clone kit.
8. Trizol reagent (*see* **Note 2**).
9. Kanamycin (*see* **Note 3**).
10. Rifampicin.
11. Restriction enzymes *Stu* I and *BamH* I-HF.
12. Agarose.
13. *TaKaRa LA Taq.*
14. Biotin-dUTP.
15. Twenty percent acrylamide/bis solution (19:1).
16. Urea.
17. Chemiluminescent nucleic acid detection module kit (including blocking buffer; stabilized streptavidin-HRP conjugate; 4 × washing buffer; substrate equilibration buffer; luminol/enhancer solution, and stable peroxide solution).
18. Perfect Hyb™ plus hybridization buffer.
19. 1-Methylimidazole.
20. 37% (v/v) formaldehyde.
21. EDC: N-(3-dimethylaminopropyl)-N′-ethyl carbondiimine hydrochloride.
22. TEMED: Tetramethylethylenediamine.
23. LB medium: Weight 10 g tryptone, 10 g sodium chloride (NaCl), and 5 g yeast extract, added deionized water to 1 L, autoclave at 121 °C for 20 min. Add 1.5% (w/v) agar for solid LB medium.
24. 10× MOPS buffer: 0.2 M 3-(*N*-morpholino) propanesulfonic acid (MOPS), 0.08 M sodium acetate, 0.01 M EDTA (pH 8.0). Adjust pH to 7.0 with NaOH. Sterilize the solution by filtration through a 0.2-micron Millipore filter. Store in the dark.
25. 20× SSC: 0.3 M trisodium citrate, 3 M NaCl, adjust pH at 7.0 with diluted HCl.

    10% (w/v) SDS.
26. Methylene blue staining solution: 0.5% (w/v) methylene blue in ddH_2O.

27. 5×TBE: Dissolve 54 g Tris-HCl (hydroxymethyl) aminomethane, 27.5 g boric acid and add 20 mL 0.5 M EDTA (pH 8.0) in 1 L ddH_2O.
28. 10% Ammonium persulfate: Dissolved 10% (w/v) ammonium persulfate in ddH_2O.
29. 2 × Small RNA loading buffer: 95% formamide, 0.025% (w/v) SDS,0.025% (w/v) bromophenol blue, 0.025% (w/v) xylene cyanol FF, 0.5 mM EDTA (pH 8.0).
30. 2 × Total RNA loading buffer: 95% formamide, 0.025% (w/v) SDS, 0.025% (w/v) bromophenol blue, 0.025% (w/v) xylene cyanol FF, 0.5 mM EDTA (pH 8.0).

2.4 Equipment

1. UVP hybridization oven.
2. Amersham lmager 600.
3. Whatman paper.
4. Electrophoresis chamber and power supply.
5. Kitchen paper.
6. Hybridization oven.
7. Amersham Hybond™—NX membrane (GE Healthcare Life Sciences) (*see* **Note 4**).
8. Amersham Hybond™—N^+ membrane semidry transfer cell.

3 Methods

3.1 Cloning of the Full-Length RTIV Sequence

1. Extract the total RNA using TRIzol according to the manufacturer's protocol (*see* **Note 5**).
2. Assemble viral sequences from wild rice RNA/sRNA high-throughput sequencing, confirm 3′ and 5′ ends by RACE, and splice to obtain RTIV full-length sequence.
3. Synthesize the first complementary DNA (cDNA) using virus gene-specific primer (GSP) according to the manufacturer's protocol.
4. Amplify the virus full length fragment with high-fidelity DNA polymerase.
5. Analyze the PCR product in 1% agarose gel by electrophoresis.
6. Excise the amplicon from the agarose gel with a clean scalpel and purify the target DNA using a DNA gel extraction kit.

3.2 Construction of Infectious cDNA Clone of pCass4-RZ-RTIV

1. Digest the pCass4-RZ plasmid with *Stu* I and *BamH* I, incubate the mixture at 37 °C for 2 h, followed by inactivation of the restriction endonuclease at 85 °C for 5 min. The digestion

products can be purified either by gel extraction or direct purification.

2. Mix 120 ng full-length viral fragment with 200 ng Stu I/BamH I-digested pCass4-Rz vector, incubate at 37 °C for 30 min in a thermal cycler, then immediately chill on ice for 5 min (*see* **Note 6**).
3. Transform ligation products into *Escherichia coli* competent cells, plate on LB agar containing kanamycin, and incubate overnight at 37 °C.
4. Pick single colonies for PCR verification and select positive bacteria for sequencing.
5. Cultivate positive colonies, then extract plasmids using the OMEGA E.Z.N.A.® Plasmid DNA Mini Kit according to the manufacturer's instructions to obtain pCass4-RZ-RTIV plasmids.

3.3 Agrobacterium-Mediated Needle Inoculation of RTIV to rice plant

1. Mix 1 μg of the constructed vector with Agrobacterium (GV3101) competent cell (200 μL) and place on ice for 5 min.
2. After freeze it in liquid nitrogen for 5 min, incubate at 37 °C for 5 min, then add LB medium, and incubate in a shaker at 28 °C for 2–3 h.
3. Pipette an appropriate amount of bacteria liquid evenly on a solid LB plate (containing the Kanamycin and Rifampicin antibiotics), then incubate at 28 °C for about 48 h.
4. Pick a single colony, verify by PCR, and expand positive clones in LB liquid medium containing kanamycin and rifampicin at 28 °C overnight.
5. Plate the cultured bacterial solution onto solid medium containing kanamycin and rifampicin, invert, and incubate at 28 °C for 3–4 days until thick bacterial growth appears (*see* **Note 7**).
6. Use five-leaf seedlings (3 weeks old) for virus inoculation (*see* **Note 8**).
7. Scratch Zhonghua-11 stems and leaves with a syringe tip, then scrape Agrobacterium from the culture medium and apply evenly to wounded tissues (*see* **Note 9**).
8. One-week post-inoculation, harvest Zhonghua-11 leaves and stem sections (*see* **Note 10**).
9. Five weeks post-inoculation, photograph plants to document symptoms. RTIV-infected plants show reduced tillering compared to mock-infected controls (Fig. 1a), and harvest newly expanded leaves for RNA extraction and viral detection.

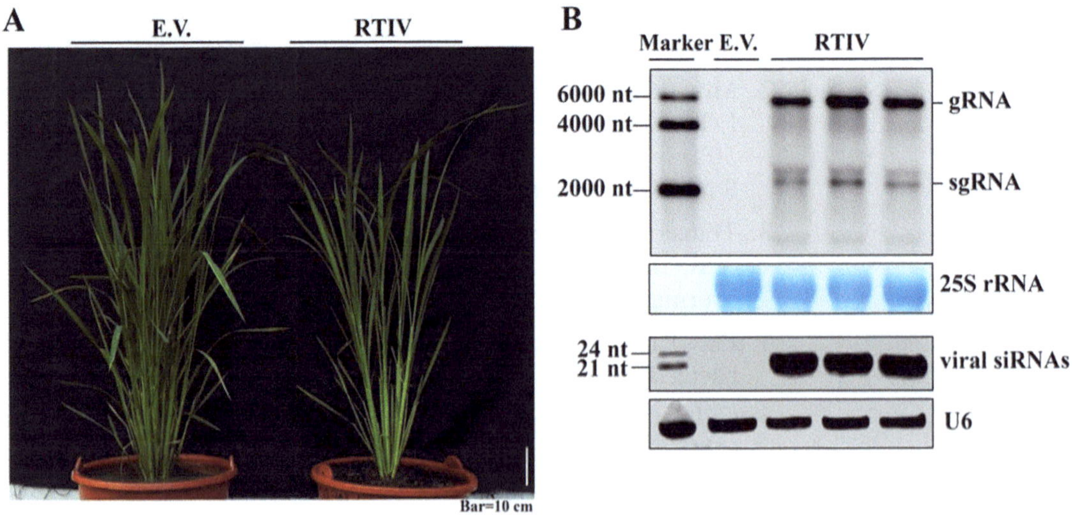

Fig. 1 Virus rescue and disease induction by an infectious cDNA clone of RTIV. (**a**) Reduced tillering phenotype in Zhonghua-11 rice plants approximately 6 weeks after agroinoculation. (**b**) Northern blot detection of RTIV genomic/subgenomic (sg) RNAs (upper panel) and RTIV-derived siRNAs (lower panel) in upper, uninoculated leaves of Zhonghua-11 rice plants. Loading controls were 25S rRNA (upper panel) and U6 RNA (lower panel)

3.4 Detection of Viral gRNA and siRNAs in Infected Rice by Northern Blot

1. Perform Northern blot analysis of RTIV genomic RNAs and siRNAs as described [19] with modifications. Separate total RNAs on formaldehyde-denaturing gels (horizontal electrophoresis) and small RNAs on UREA-denaturing gels (vertical electrophoresis).
2. Transfer total RNAs to N-Plus (+) membrane by capillary blotting and small RNAs to NX membrane by semi-dry transfer. Pre-hybridize membranes for 2 h at 65 °C (gRNAs) or 42 °C (vsiRNAs) in 20 mL pre-warmed hybridization buffer (*see* **Note 11**).
3. Prepare biotin-labeled probes using TaKaRa LA Taq and Biotin-dUTP. PCR mixture (20 μL total): 2 μL 10× LA PCR Buffer II, 2 μL dNTP mix (A/C/G, 2.5 mM each), 0.35 μL dUTP (2.5 mM), 0.65 μL Biotin-11-dUTP (1 mM), 0.8 μL CP-probe-F primer (10 μM), 0.8 μL CP-probe-R primer (10 μM), 0.35 μL LA Taq (5 U/μL), 0.3 μL template DNA, and ddH_2O to 20 μL. Primers: CP-probe-F (5′-CAATGCGCT GACGATCAATTG-3′) and CP-probe-R (5′-GGTGACGTCG GAAATCAACCAC-3′).
4. Add denatured biotinylated probe (~30 ng/mL hybridization buffer) after pre-hybridization (*see* **Note 12**).
5. Hybridize overnight at 65 °C (gRNAs) or 42 °C (vsiRNAs) with rotation (5 rpm).

6. Wash membranes sequentially: 1× SSC/0.1% SDS for 10 min, 0.2× SSC/0.1% SDS for 10 min, and 0.1× SSC/0.1% SDS for 5 min (27 °C each) (*see* **Note 13**).
7. Block membrane with blocking buffer (pre-warmed to 37 °C) for 15 min at room temperature. Incubate with stabilized streptavidin–horseradish peroxidase conjugate (1:300 in blocking buffer) for 15 min at room temperature. Wash four times with 1× wash buffer (5 min each) at room temperature. Incubate with substrate equilibration buffer for 5 min at room temperature (*see* **Note 13**).
8. Prepare substrate working solution (1:1 luminol/enhancer solution and stable peroxide solution) to cover membrane completely (*see* **Note 14**).
9. Expose membrane to Amersham Imager 600, adjusting exposure time as needed. RTIV genomic RNAs and vsiRNAs are detected in systemic leaves at 6 weeks post-inoculation (Fig. 1b).

4 Notes

1. For more details about the plasmids pCass4-RZ, refer to [20].
2. All toxic reagents must be operated within a fume hood to prevent leakage that are harmful to health and must be used in strict accordance with the manufacturer's instructions (including DEPC, EDC, TRIzol, etc.).
3. All chemicals are molecular biology-grade chemicals. Store all buffers at appropriate temperature (unless indicated otherwise).
4. All membranes need to be soaked and moistened in the corresponding transfer buffer for about 5 min before transfer (including N^+ Membrane, NX membrane).
5. Compared to using total RNA from wild rice as the template for reverse transcription, it is recommended to use RNA extracted from enriched viral particles as the template, which facilitates more efficient amplification of the virus full-length sequence.
6. The pCass4-Rz vector is selected as the infectious clone vector due to its suitability for generating transgenic rice lines, thereby facilitating subsequent investigation of virus–rice interactions.
7. The Agrobacterium culture must form a dense bacterial lawn on the Petri dish to facilitate subsequent inoculation and infection.
8. Plants should not be allowed to become overly mature to avoid unsuccessful infection.

9. It is recommended to create mild wounds on rice stems and leaves using a syringe needle to facilitate *Agrobacterium* infection. Excessive force should be avoided during wounding to prevent severe tissue damage, which may compromise infection efficiency. Compared to leaves, inoculating the stem of rice enables *Agrobacterium* to infect the host more efficiently through the phloem tissues.
10. It is necessary to remove the inoculated leaves and stems, which can prevent *Agrobacterium* contamination during subsequent virus detection.
11. Before pre-hybridization, pre-warm hybridization buffer, tubes, and oven at 65 °C for gRNAs detection or 42 °C for vsiRNAs detection at least half hour.
12. While pre-hybridizing, denature the biotinylated DNA probe. Heat DNA probe at 100 °C for 5 min and place on ice for 5 min.
13. It is recommended to use the high rotation speed during washing steps, and a low rotation speed during incubation with probes, blocking buffers and substrate equilibration buffer.
14. The working solution should be kept in an amber bottle and avoid prolonged exposure to any intense light.

Acknowledgments

The authors are indebted to Prof. Shou-wei Ding (University of California, Riverside) for the pCass4-RZ vector. This research was funded by grants from the National Natural Science Foundation of China (32402389) and China Postdoctoral Science Foundation (2024 M761532) to W.Y., the Fujian Hundred Talent grant to Z.G.

References

1. Elert E (2014) Rice by the numbers: a good grain. Nature 514:S50–S51
2. Hibino H (1996) Biology and epidemiology of rice viruses. Annu Rev Phytopathol 34:249–274
3. Wu JG, Yang GY, Zhao SS et al (2022) Current rice production is highly vulnerable to insect-borne viral diseases. Natl Sci Rev 9:nwac131
4. Li L, Chen J, Sun Z (2024) Exploring the shared pathogenic strategies of independently evolved effectors across distinct plant viruses. Trends Microbiol 32:1021–1033
5. Wei J, Jia D, Mao Q et al (2018) Complex interactions between insect-borne rice viruses and their vectors. Curr Opin Virol 33:18–23
6. Iqbal Z, Sattar MN, Naqqash MN (2022) Diversity and management of plant viruses infecting rice. In: Sarwar N, Atique-ur-Rehman, Ahmad S, Hasanuzzaman M (eds) Modern techniques of rice crop production. Springer, Singapore
7. Feng M, Li L, Cheng R et al (2021) Development of a mini-replicon-based reverse-genetics system for rice stripe Tenuivirus. J Virol 95: e0058921

8. Shakir S, Zaidi SS, Hashemi FSG et al (2023) Harnessing plant viruses in the metagenomics era: from the development of infectious clones to applications. Trends Plant Sci 28:297–311
9. Brewer HC, Hird DL, Bailey AM (2018) A guide to the contained use of plant virus infectious clones. Plant Biotechnol J 16:832–843
10. Abrahamian P, Hammond RW, Hammond J (2020) Plant virus-derived vectors: applications in agricultural and medical biotechnology. Annu Rev Virol 7:513–535
11. Bisaro DM, Hamilton WD, Coutts RH et al (1982) Molecular cloning and characterisation of the two DNA components of tomato golden mosaic virus. Nucleic Acids Res 10:4913–4922
12. Ahlquist P, French R, Janda M (1984) Multicomponent RNA plant virus infection derived from cloned viral cDNA. Proc Natl Acad Sci USA 81:7066–7070
13. Ganesan U, Bragg JN, Deng M et al (2013) Construction of a sonchus yellow net virus minireplicon: a step toward reverse genetic analysis of plant negative-strand RNA viruses. J Virol 87:10598–10611
14. Wang Q, Ma X, Qian S et al (2015) Rescue of a plant negative-strand RNA virus from cloned cDNA: insights into enveloped plant virus movement and morphogenesis. PLoS Pathog 11:e1005223
15. Feng M, Cheng R, Chen M et al (2020) Rescue of tomato spotted wilt virus entirely from complementary DNA clones. Proc Natl Acad Sci USA 117:1181–1190
16. Wu Q, Ding SW, Zhang Y et al (2015) Identification of viruses and viroids by next-generation sequencing and homology-dependent and homology-independent algorithms. Annu Rev Phytopathol 53:425–444
17. Wu Q, Luo Y, Lu R et al (2010) Virus discovery by deep sequencing and assembly of virus-derived small silencing RNAs. Proc Natl Acad Sci USA 107:1606–1611
18. Yan W, Zhu Y, Liu W et al (2023) Discovery of aphid-transmitted Rice tiller inhibition virus from native plants through metagenomic sequencing. PLoS Pathog 19:e1011238
19. Guo Z, Wang XB, Li WX et al (2019) A sensitized genetic screen to identify novel components and regulators of the host antiviral RNA interference pathway. Methods Mol Biol 2028:215–229
20. Annamalai P, Rao AL (2005) Replication-independent expression of genome components and capsid protein of brome mosaic virus in planta: a functional role for viral replicase in RNA packaging. Virology 338:96–111

Chapter 20

Methods for Autophagy Detection by Fluorescence Microscopy

Xiaoye Ke, Xianglong Zhang, and Xiangxiang Zhang

Abstract

Fluorescence microscopy is pivotal for investigating autophagy's role in plant antiviral immunity. Here, we present a standardized procedure using complementary probes, CFP-ATG8f for autophagosomal structures and monodansylcadaverine (MDC) for autophagic vacuoles, to assess autophagy during viral infection. This combined CFP-ATG8f and MDC staining system provides a powerful, reproducible method for evaluating autophagic activity in plant–virus interactions.

Key words Autophagy, CFP-ATG8f, MDC staining, Confocal microscopy

1 Introduction

Autophagy is an evolutionarily conserved degradation mechanism that involves the sequestration of cytoplasmic components within double-membraned autophagosomes and their subsequent delivery to the vacuole or lysosome for breakdown and recycling [1]. This process is governed by a suite of autophagy-related (ATG) genes encoding core molecular components. In plants, autophagy plays vital roles in development, reproduction, metabolism, senescence, and adaptation to abiotic and biotic stresses [2, 3]. Notably, autophagy is induced upon viral infection, with accumulating evidence underscoring its contribution to antiviral defense during plant–virus interactions [4–6]. For example, autophagy is upregulated during the resistance response to tobacco mosaic virus (TMV), and plants impaired in ATG function exhibit enhanced viral accumulation, highlighting its importance in restricting viral infection [7]. Given the critical function of autophagy in plant antiviral immunity, there is a pressing need for robust labeling techniques to accurately quantify autophagic activity.

Fluorescence-based approaches leverage multiple specialized probes to monitor distinct stages of autophagy. The CFP–ATG8

Guanwei Wu and Xiaofei Cheng (eds.), *Plant Antiviral Resistance: Methods and Protocols*, Methods in Molecular Biology, vol. 3039, https://doi.org/10.1007/978-1-0716-5300-5_20,

fusion protein serves as a specific marker for autophagosome formation, enabling quantitative tracking of ATG8-positive puncta formation [8]. Meanwhile, monodansylcadaverine (MDC), an auto-fluorescent dye, selectively labels autophagic vacuoles and offers complementary information on autophagosome maturation [9]. When integrated with advanced imaging platforms such as confocal microscopy, these probes provide exceptional spatiotemporal resolution for dissecting autophagic dynamics [10].

In antiviral research, fluorescence microscopy has proven particularly valuable. Live-cell imaging of CFP-ATG8 allows direct observation of autophagic flux during viral infection, while MDC staining facilitates assessment of autophagic vesicle accumulation [11, 12]. These tools are instrumental in elucidating the dual role of autophagy, acting both as an antiviral mechanism and, in certain contexts, as a pathway exploited by viruses to facilitate their replication [13, 14]. Here, we describe a standardized dual-labeling protocol employing CFP-ATG8f as an autophagosome marker and monodansylcadaverine (MDC) to stain autophagic vacuoles, enabling monitoring of autophagy during viral infection. The combined CFP-ATG8f and MDC system provides a robust and reproducible approach for evaluating autophagy in plant–virus interactions.

2 Materials

2.1 Plants, Bacterial Strains and Plasmids

1. *Nicotiana benthamiana* seeds.
2. *Escherichia coli* strains DH5α competent cells.
3. *Agrobacterium tumefaciens* strain GV3101.
4. pCV-CFP and pCV-RFP plasmids were kindly provided by Prof. Fei Yan from Ningbo University.

2.2 Equipment

1. Thermostatic incubator.
2. Autoclave machine.
3. 50 mL Falcon tubes.
4. pH meter.
5. 0.22 μm syringe filters.
6. Petri dishes.
7. Pipettes and tips.
8. 1.5 mL and 2.0 mL Eppendorf tubes.
9. Eppendorf centrifuge 5420.
10. Vortex oscillator.
11. GenePulser/MicroPulser Electroporation Cuvettes, 0.2 cm gap.

12. Eppendorf BioPhotometer.
13. Electrophoresis equipment.
14. Confocal microscopy (Leica TCS SP8, Mannheim, Germany).
15. Slides and coverslips.
16. Plant leaf sampler and tweezers.

2.3 Regents and Buffers

1. LB liquid medium: dissolve 5 g of yeast extract, 10 g of peptone, and 10 g of NaCl in 800 mL of distilled water. Adjust pH to 7.0 with NaOH, bring volume to 1 L with distilled water, and sterilize by autoclaving at 121 °C for 20 min.
2. LB solid medium: add 20 g of agar to 1 L LB liquid medium and sterilize the mixture by autoclaving.
3. 0.5 M EDTA solution (pH 8.0): Dissolve 18.61 g EDTA disodium salt in ~80 mL distilled water. Adjust pH to 8.0 with NaOH, and bring volume to 100 mL with distilled water.
4. 50 × TAE (Tris–acetate–EDTA) buffer: Dissolve 242 g Tris base in ~700 mL distilled water. Add 57.1 mL glacial acetic acid and 100 mL of 0.5 M EDTA solution (pH 8.0). Bring volume to 1 L with distilled water. Prepare 1× working solution by diluting stock 1:50.
5. Agrobacterium infiltration buffer: mix 5 mL 1 M $MgCl_2$, 5 mL 1 M MES, and 500 μL 200 mM acetosyringone (AS). Bring volume to 500 mL with sterile distilled water. Mix thoroughly and store at 4 °C until use.
6. 50 mM monodansylcadaverine (MDC) stock solution: dissolve MDC in anhydrous dimethyl sulfoxide (DMSO), mix thoroughly, aliquot into single-use volumes, and store protected from light at −20 °C.
7. 1 × phosphate buffered saline (PBS): Weigh 8 g NaCl, 0.44 g Na_2HPO_4, 0.24 g KH_2PO_4, 0.2 g KCl, dissolve them in ~800 mL ddH_2O, adjust the pH of solution to 7.5 using NaOH, and bring the final volume to 1 L.
8. PrimeSTAR GXL Premix (TAKARA).
9. OMEGA Gel Extraction Kit.
10. T4 DNA polymerase.
11. dATP, dTTP.
12. Fast Digest ApaI.
13. Kanamycin.
14. OMEGA Plasmid Mini Kit I.
15. E-64d (Sigma).

3 Methods

3.1 Construction of a CFP-NbATG8f Expression Vector Using Ligation-Independent Cloning

1. Design primers specific for NbATG8f coding region (GenBank: KU561372).
2. Amplify *NbATG8f* by using PrimeSTAR GXL Premix: mix 15 μL 2 × Premix, 2 μL each primer (10 uM), 2 μL cDNA (100 ng), and 9 μL ddH_2O (30 μL total). Cycle: 98 °C 2 min; 30–35 cycles (98 °C 20 s, 58 °C 20 s, 68 °C 1 min/kb); 68 °C 7 min.
3. Mix PCR products with 5 μL 5× DNA loading buffer. Separate on 1% agarose gel at 180 V for 20 min. Verify band size under UV against DNA marker.
4. Excise target bands under UV with scalpel. Purify DNA using OMEGA Gel Extraction Kit.
5. Set up dA-tailing: 5 μL purified PCR product, 1 μL 10× buffer, 0.5 μL 100 nM dATP, 0.2 μL T4 DNA Polymerase, 3.3 μL ddH_2O (10 μL total). Incubate at 37 °C 20 min, then 75 °C 20 min.
6. Linearize pCV-CFP vector with Fast Digest ApaI per manufacturer's protocol.
7. Purify digested vector using OMEGA Gel Extraction Kit.
8. Set up dT-tailing: 30 μL linearized vector, 10 μL 10× buffer, 10 μL 100 nM dTTP, 5 μL T4 DNA Polymerase, 45 μL ddH_2O (100 μL total). Incubate at 37 °C 20 min, then 75 °C 20 min.
9. Mix equal volumes (2 μL each) of dA-tailed insert and dT-tailed vector. Anneal 70 °C 20 min, then 22 °C 20 min.
10. Transform entire annealing product into *E. coli DH5α* competent cells. Plate on LB agar with kanamycin. Incubate overnight at 37 °C.
11. Screen single colonies by PCR and confirm positive clones by DNA sequencing. 12. Culture positive clones in liquid medium. Extract CFP-ATG8f plasmid using OMEGA Plasmid Mini Kit I per manufacturer's instructions.

3.2 Agroinfiltration--Mediated Transient Expression in Plant Epidermal Cells

1. Introduce CFP-NbATG8f plasmid into *A. tumefaciens* GV3101 by electroporation. Plate on LB agar with appropriate antibiotics and incubate at 28 °C for 36–48 h.
2. Select colonies for PCR verification and DNA sequencing. Inoculate positive clones into LB liquid medium with corresponding antibiotics and grow overnight at 28 °C, 180 rpm.
3. Centrifuge Agrobacterial cultures at 7000 rpm for 2 min at room temperature.

4. Discard supernatant and resuspend pellet in agrobacterium infiltration buffer. Repeat harvest and wash steps twice.
5. Dilute washed Agrobacterial suspensions to final OD600 = 0.3 with agrobacterium infiltration buffer.
6. Infiltrate Agrobacterial suspensions into *N. benthamiana* leaves using a 1-mL needleless syringe.

3.3 CFP-NbATG8f Labeling of Autophagosomes

1. Co-express CFP-NbATG8f marker with test plasmids or virus infectious clone in five to six-leaf stage *N. benthamiana* leaves via agroinfiltration (*see* **Notes 1–3**).
2. At 48 h post infiltration (hpi), infiltrate leaves with 100 mM E-64d and incubate in dark for 8–12 h (*see* **Note 4**).
3. Observe samples by confocal microscopy (excitation 405 nm; emission 454–475 nm; *see* **Notes 5** and **6**).
4. Expected results: Use NbATG5 (autophagy activator) as positive control. ATG5-RFP treatment shows increased cyan fluorescent puncta vs. RFP control, confirming ATG5 activation of autophagy (Fig. 1a).

3.4 MDC Staining of Autophagosomes

1. Infiltrate leaves with 100 mM E-64d and incubate in dark for 8–12 h.
2. Excise leaf sections and vacuum-infiltrate with 50 mM MDC for 8–14 min.
3. Rinse samples twice with 1 × PBS (*see* **Note** 7).
4. Punch 0.5 cm × 0.5 cm leaf discs, mount on slides, and image by confocal microscopy with a 40× oil-immersion objective (MDC: excitation 335 nm, emission 450–550 nm; chlorophyll: excitation 543 nm, emission 580–700 nm) (*see* **Note 8**).
5. Expected results: Use NbATG5 as positive control. ATG5-RFP treatment shows increased green fluorescent puncta vs. RFP control, confirming MDC as a reliable autophagosome marker (Fig. 1b).

4 Notes

1. The optical density of the Agrobacterium culture used for infiltration should be set at an OD_{600} of 0.2–0.3; higher concentrations must be avoided to prevent potential toxicity.
2. Plants should be maintained under adequate watering conditions prior to inoculation to avoid drought or water deprivation stress.
3. The pH of Agrobacterium infiltration buffer critically influences infection efficiency and should be carefully adjusted.

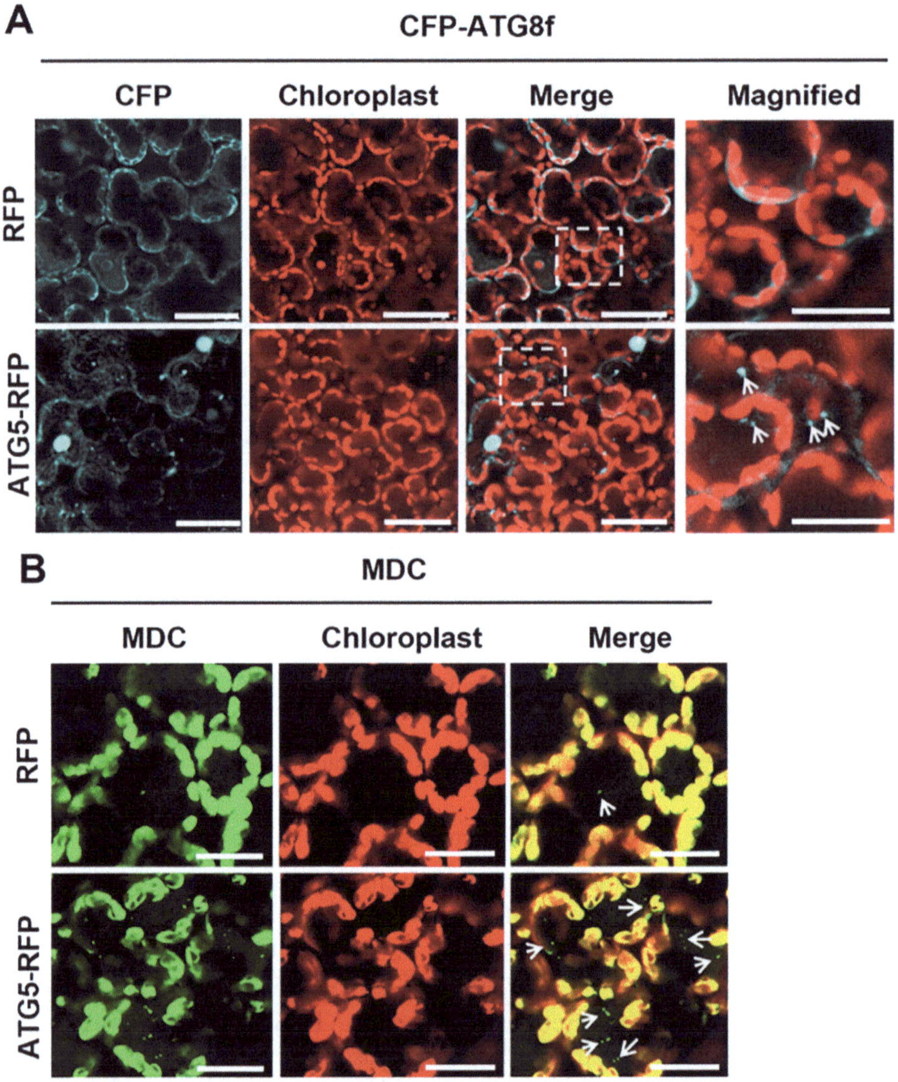

Fig. 1 Autophagosome detection in *N. benthamiana*. (**a**) CFP-NbATG8f reporter. (**b**) MDC staining. Leaves were agroinfiltrated with RFP or NbATG5-RFP and imaged by confocal microscopy. Magnified regions (white dashed boxes) and representative autophagosomes (white arrows) are indicated. Scale bars: 20 μm

4. Treat leaves by infiltration with 100 mM E-64d (Sigma), an autophagy inhibitor, and incubate them in the dark for 8–12 hpi.
5. When imaging multiple fluorophores, the sequential scanning mode should be used to minimize signal crosstalk.
6. Activate RFP channel to visualize chloroplast morphology. Focus on lower epidermal cells where autophagosomes reside within chloroplast-surrounded vacuoles.

7. Rinse MDC-stained samples twice with 1× PBS to remove surface dye. This washing step is critical to minimize background fluorescence that could obscure autophagosome detection.
8. For MDC detection (excitation 335 nm, emission 450–550 nm), increase laser intensity to 2× standard level. Focus on green fluorescent puncta within chloroplast-surrounded vacuoles, as MDC selectively stains autophagosomes in the vacuolar lumen.

Acknowledgments

The authors gratefully acknowledge Prof. Fei Yan (Ningbo University) for the generous gift of pCV-CFP and pCV-RFP plasmids. This work was supported by grants from the National Natural Science Foundation of China (32402302) to XZ.

References

1. He C, Klionsky DJ (2009) Regulation mechanisms and signaling pathways of autophagy. Annu Rev Genet 43:67–93
2. Zhou J, Yu JQ, Chen Z (2014) The perplexing role of autophagy in plants innate immune responses. Mol Plant Pathol 15:637–645
3. Hofius D, Li L, Hafren A, Coll NS (2017) Autophagy as an emerging arena for plant-pathogen interactions. Curr Opin Plant Biol 38:117–123
4. Leary AY, Savage Z, Tumtas Y, Bozkurt TO (2019) Contrasting and emerging roles of autophagy in plant immunity. Curr Opin Plant Biol 52:46–53
5. Yang M, Ismayil A, Liu Y (2020) Autophagy in plant-virus interactions. Annu Rev Virol 7: 403–419
6. Ismayil A, Yang M, Liu Y (2020) Role of autophagy during plant-virus interactions. Semin Cell Dev Biol 101:36–40
7. Liu Y, Schiff M, Czymmek K, Talloczy Z, Levine B, Dinesh-Kumar SP (2005) Autophagy regulates programmed cell death during the plant innate immune response. Cell 121: 567–577
8. Contento AL, Bassham DC (2012) Structure and function of the ATG8 conjugation system in autophagy. Curr Opin Plant Biol 15:721–729
9. Biederbick A, Kern HF, Elsässer HP (1995) Monodansylcadaverine (MDC) is a specific in vivo marker for autophagic vacuoles. Eur J Cell Biol 66:3–14
10. Ismayil A, Yang M, Haxim Y et al (2020) Cotton leaf curl Multan virus βC1 protein induces autophagy by disrupting the interaction of autophagy related protein 3 with glyceraldehyde-3 phosphate dehydrogenases. Plant Cell 32:1124–1135
11. Vicinanza M, Gratian MJ, Bowen M, Rubinsztein DC (2017) Fluorescence-based assays to analyse phosphatidylinositol 5-phosphate in autophagy. Methods Enzymol 587:311–330
12. Xu GY, Wang SS, Han SJ et al (2017) Plant bax inhibitor-1 interacts with ATG6 to regulate autophagy and programmed cell death. Autophagy 13:1161–1175
13. Li ZM, Duan SH, Yu TM et al (2024) Bunyavirus SFTSV NSs utilizes autophagy to escape the antiviral innate immune response. Autophagy 20:2133–2145
14. Zhang XX, Yin YY, Su YH et al (2021) eIF4A, a target of siRNA derived from rice stripe virus, negatively regulates antiviral autophagy by interacting with ATG5 in *Nicotiana benthamiana*. PLoS Pathog 17:e1009963

Index

Guanwei Wu and Xiaofei Cheng (eds.), *Plant Antiviral Resistance: Methods and Protocols*, Methods in Molecular Biology, vol. 3039, https://doi.org/10.1007/978-1-0716-5300-5,

MIX
Papier aus verantwortungsvollen Quellen
Paper from responsible sources
FSC® C105338

If you have any concerns about our products,
you can contact us on
ProductSafety@springernature.com

In case Publisher is established outside the EU,
the EU authorized representative is:
Springer Nature Customer Service Center GmbH
Europaplatz 3, 69115 Heidelberg, Germany

Printed by Libri Plureos GmbH
in Hamburg, Germany